T0252628

Wax Deposition

Experimental Characterizations,
Theoretical Modeling, and Field Practices

Emerging Trends and Technologies in Petroleum Engineering

Series Editor
Abhijit Y. Dandekar

PUBLISHED TITLES:

Wax Deposition: Experimental Characterizations, Theoretical Modeling, and Field Practices, *Zhenyu Huang, Sheng Zheng, H. Scott Fogler*

Wax Deposition

Experimental Characterizations, Theoretical Modeling, and Field Practices

Zhenyu Huang
Sheng Zheng
H. Scott Fogler

CRC Press
Taylor & Francis Group
Boca Raton   London   New York

CRC Press is an imprint of the
Taylor & Francis Group, an **informa** business

CRC Press
Taylor & Francis Group
6000 Broken Sound Parkway NW, Suite 300
Boca Raton, FL 33487-2742

First issued in paperback 2020

© 2015 by Taylor & Francis Group, LLC
CRC Press is an imprint of Taylor & Francis Group, an Informa business

No claim to original U.S. Government works

ISBN-13: 978-1-4665-6766-5 (hbk)
ISBN-13: 978-0-367-78349-5 (pbk)

This book contains information obtained from authentic and highly regarded sources. Reasonable efforts have been made to publish reliable data and information, but the author and publisher cannot assume responsibility for the validity of all materials or the consequences of their use. The authors and publishers have attempted to trace the copyright holders of all material reproduced in this publication and apologize to copyright holders if permission to publish in this form has not been obtained. If any copyright material has not been acknowledged please write and let us know so we may rectify in any future reprint.

Except as permitted under U.S. Copyright Law, no part of this book may be reprinted, reproduced, transmitted, or utilized in any form by any electronic, mechanical, or other means, now known or hereafter invented, including photocopying, microfilming, and recording, or in any information storage or retrieval system, without written permission from the publishers.

For permission to photocopy or use material electronically from this work, please access www.copyright.com (http://www.copyright.com/) or contact the Copyright Clearance Center, Inc. (CCC), 222 Rosewood Drive, Danvers, MA 01923, 978-750-8400. CCC is a not-for-profit organization that provides licenses and registration for a variety of users. For organizations that have been granted a photocopy license by the CCC, a separate system of payment has been arranged.

Trademark Notice: Product or corporate names may be trademarks or registered trademarks, and are used only for identification and explanation without intent to infringe.

Visit the Taylor & Francis Web site at
http://www.taylorandfrancis.com

and the CRC Press Web site at
http://www.crcpress.com

Contents

Series Preface

This petroleum engineering book series includes works on all aspects of petroleum science and engineering but with special focus on emerging trends and technologies that pertain to the paradigm shift in the petroleum engineering field. It deals with the increased exploitation of technically challenged and atypical hydrocarbon resources that are receiving a lot of attention from today's petroleum industry, as well as the potential use of advanced nontraditional or nonconventional technologies such as nanotechnology in diverse petroleum engineering applications. These areas have assumed a position of prominence in today's petroleum engineering field. However, although scientific literature exists on these emerging areas in the form of various publications, much of it is scattered and highly specific. The purpose of this book series is to provide a centralized and comprehensive collection of reference books and textbooks covering the fundamentals but paying close attention to these emerging trends and technologies from the standpoint of the main disciplines of drilling engineering, production engineering, and reservoir engineering.

Given the dwindling supply of easy-to-produce conventional oil, rapidly climbing energy demands, the sustained ~$100/bbl. oil price, and technological advances, the petroleum industry is increasingly in pursuit of exploration and production (E&P) of atypical or unconventional and technically challenged oil and gas resources, which may eventually become the future of the petroleum industry. Unconventional resources typically include (1) coal bed methane (CBM) gas; (2) tight gas in ultralow permeability formations; (3) shale gas and shale oil in very low permeability shales; (4) oil shales; (5) heavy and viscous oils; (6) tar sands; and (7) methane hydrates. Compared to the world's proven conventional natural gas reserves of 6600+ trillion cubic feet (TCF), the combined CBM, shale gas, tight gas, and methane hydrate resource estimates are in excess of 730,000 TCF.[1-3] Similarly, out of the world's total of 9 to 13 trillion bbl. of oil resources, the conventional (light and medium oil) is only 30%, whereas heavy oil, extra-heavy oil, tar sands, and bitumen combined make up the remaining 70%.[4] In addition shale-based oil resources worldwide are estimated to be between 6 and 8 trillion bbl.[5] As a case in point, shale-based oil production in North Dakota has increased from a mere 3000 bbl./day in 2005 to a whopping 400,000+ bbl./day in 2011.[6] Even the most conservative technical and economic recovery estimates of the unconventional resources represent a very substantial future energy portfolio that dwarfs the conventional gas and oil reserves. However, to a large extent, these particular resources, unlike the conventional ones, do not fit the typical profile and are to some extent in the stages of infancy, thus needing a different and unique approach from the drilling, production, and reservoir engineering perspectives.

The petroleum engineering academic and industry community is also aggressively pursuing nanotechnology with the hope of identifying innovative solutions for problems faced in the overall process of oil and gas recovery. In particular, a big spurt in this area in the last decade or so is evident from the significant activities in terms of research publications, meetings, formation of different consortia, workshops, and dedicated sessions in petroleum engineering conferences. A simple literature search for a keyword *nanotechnology* on http://www.onepetro.org, managed by the Society of Petroleum Engineers (SPE), returns over 250 publications dating from 2001 onward with the bulk of them in the last 5 or 6 years. Since 2008, SPE also organized three different applied technology workshops specifically focused on nanotechnology in the E&P industry. An Advanced Energy Consortium with sponsorships from some major operators and service companies was also formed in 2007 with the mission of facilitating research in "micro and nanotechnology materials and sensors having the potential to create a positive and disruptive change in the recovery of petroleum and gas from new and existing reservoirs." Companies such as Saudi Aramco have taken the lead in taking the first strides in evaluating the potential of employing nanotechnology in the E&P industry. Their trademarked Resbots™ are designed for deployment with the injection fluids for in situ reservoir sensing (temperature, pressure, and fluid type) and intervention, eventually leading to more accurate reservoir characterization once fully developed. Following successful laboratory core flood tests, they conducted the industry's first field trial of reservoir nanoagents.[7]

The foregoing is clearly a statement of the new wave in the petroleum engineering field, which is being created by emerging trends in unconventional resources and new technologies. The publisher and its series editor are fully aware of the rapidly evolving nature of these key areas and their long-lasting influence on the current state and future of the petroleum industry. The series is envisioned to have a very broad scope that includes but is not limited to analytical, experimental, and numerical studies and methods and field cases, thus delivering readers in both academia and industry an authoritative information source of trends and technologies that have shaped and will continue to impact the petroleum industry.

References

1. Retrieved from http://www.eia.gov/analysis/studies/worldshalegas/ (accessed date June 10, 2013).

2. Kawata, Y. & Fujita, K. Some predictions of possible unconventional hydrocarbons availability until 2100. Society of Petroleum Engineers (SPE) paper number 68755. SPE Asia Pacific Oil and Gas Conference and Exhibition, 17–19 April, Jakarta, Indonesia.
3. Retrieved from http://www.netl.doe.gov/kmd/cds/disk10/collett.pdf. Methane Hydrates Interagency R&D Conference, 20–22 March 2002, Washington, DC.
4. Retrieved from https://www.slb.com/~/media/Files/resources/oilfield_review /ors06/sum06/heavy_oil.ashx
5. Biglarbigi, K., Crawford, P., Carolus, M. & Dean, C. Rethinking world oil–shale resource estimates. Society of Petroleum Engineers (SPE) paper number SPE 135453. SPE Annual Technical Conference and Exhibition, 19–22 September, Florence, Italy.
6. Mason, J. Retrieved from http://www.sbpipeline.com/images/pdf/Mason _Oil%20Production%20Potential%20of%20the%20North%20Dakota%20 Bakken_OGJ%20Article_10%20February%202012.pdf
7. Kanj, M. Y., Rashid, M.H. & Giannelis, E.P. Industry first field trial of reservoir nanoagents. Society of Petroleum Engineers (SPE) paper number SPE 142592. SPE Middle East Oil and Gas Show and Conference, 25–28 September, Manama, Bahrain.

Abhijit Dandekar
University of Alaska Fairbanks

Preface

Wax deposition has become one of the most common flow assurance problems in the petroleum industry. As petroleum resources shift from onshore reservoirs toward offshore subsea production, the industry is currently facing unprecedented challenges to maintain flow assurance for petroleum production, in which the strategy to prevent or mitigate wax deposition has become increasingly costly and complicated.

The goal to manage the issue of wax deposition involves answering the following three questions:

- Do we have a problem for this field?
- If yes, what kind of a problem is it? How bad is it?
- How can we solve this problem?

They are typical questions to be answered not only for wax-related issues but also for a variety of many other common production chemistry problems in general flow assurance practices. The answer to the first question generally involves only a few fluid testing procedures that are relatively simple, while much more understandings on production chemistry and fluid flow are required to address the second question. The answer to the third question requires not only knowledge that are shown in chemical engineering textbooks but also significant operational experience, and the decision makers have to fully understand the production capability of the field and the effectiveness as well as implication of any mitigation/remediation methods.

While there are several books that provide general knowledge on flow assurance, a book that specifically addresses the issue of wax deposition is still not yet available. This book is the first one that covers the entire spectrum of knowledge on wax deposition phenomena. It provides a detailed description of the thermodynamic and transport theories for wax deposition modeling and a comprehensive review of the laboratory testing to help establish appropriate control strategies for the field. It provides a progressive introduction to help flow assurance engineers to understand the process of wax deposition, to be familiar with the various methods to identify its severity, and to eventually control this problem. For engineering students, practicing engineers, and researchers in the field of flow assurance, this book serves as an in-depth discussion of how fundamental principles of thermodynamics, heat, and mass transfer can be applied to solve a problem common to the petroleum industry.

Going back to the three key questions that were raised earlier in this Preface, we hope to provide valuable information in this book that could help the readers to address these questions. Chapter 1 presents the background of wax deposition, including the cause of the phenomena, the magnitude of wax deposition problems, as well as its impact on petroleum production. Chapters 2 and 3 introduce various laboratory techniques and theoretical models. These testing and modeling are indispensable to address the first question (Do we have a wax problem?). Chapters 4–6 present the knowledge that is critical to answer the second key question (How bad is the problem?). In Chapter 4, a systematic presentation will be made to describe the process of wax deposition using chemical engineering fundamentals. It discusses various models of wax deposition and analyzes the differences between the assumptions used in these models. In addition, the advantages and disadvantages in each model are compared. Chapter 5 provides a detailed description of how to conduct laboratory wax deposition experiments in order to benchmark different wax deposition models. In this chapter, the applications of the cold finger apparatus and the lab scale flow loop are highlighted. Chapter 6 discusses examples of how fundamental principles of heat and mass transfer can be applied to interpret laboratory wax deposition experiments to better understand wax deposition behaviors and eventually predict the wax deposit growth in field operations. Chapter 7 brings the readers to the "real world" by providing several field examples of how management strategies for wax deposition in the field can be established based on the available laboratory testing and modeling work, thereby addressing the third question (How can we solve the problem?).

This book contains comprehensive knowledge of wax deposition not only from academic research but also from the flow assurance industry, thanks to the comments and suggestions from many petroleum companies in the industry. We acknowledge Tommy Golczynski and Tony Spratt from Assured Flow Solutions LLC for carefully reviewing the drafts of the book. We thank all the sponsors of the University of Michigan Industrial Affiliates Program, including Chevron, ConocoPhillips, Multichem, Nalco, Shell, Statoil, Total, and Wood Group Kenny. In addition to their financial support to the academic research to the Michigan Industrial Affiliates Research Program, the expertise and experience shared by the representatives of these companies constitute an integral part of the completion of this book. We also thank all the members of Professor Fogler's research group for their effort dedicated to the Michigan Industrial Affiliates Research Program. Finally, we extend our gratitude to our family members for their support in completing this book.

Authors

 **Dr. Zhenyu Huang (Jason)** is currently a senior flow assurance specialist in Assured Flow Solutions LLC, providing engineering solutions for a variety of flow assurance issues to the petroleum industry. His expertise includes production chemistry and multiphase flows. Dr. Huang has more than 8 years of academic and industrial experience focused extensively on a variety of wax deposition problems. His work includes model development, experimental verifications, fluid testing and field applications. He has been involved with multiple offshore developments that present wax deposition/gelation concerns. He is the subject matter expert on wax-related issues, and he currently serves as the vice president of the Upstream Engineering and Flow Assurance Section of the American Institute of Chemical Engineers. Dr. Huang earned his bachelor's degree in Tsinghua University in Beijing, China, in 2006, and he completed his PhD study at the University of Michigan, Ann Arbor in 2011 with a thesis entitled "Application of the fundamentals of heat and mass transfer to the investigation of wax deposition in subsea pipelines."

 **Sheng Zheng (Mark)** graduated *summa cum laude* from the University of Michigan with a bachelor's degree in chemical engineering and minors in chemistry and mathematics. He is currently a doctoral candidate in Professor Fogler's research group, specializing in both cutting-edge experimental characterizations and theoretical modeling for wax deposition research. He has multiple high-quality publications focusing on compositional wax deposition modeling and wax transport in multiphase flow conditions. Together with Dr. Huang during their work at Wood Group Kenny, Mark carried out the DeepStar Wax Prediction and Pigging Design project, a joint industrial project to comprehensively survey and assess current industrial wax management and control strategies.

 **Dr. H. Scott Fogler** is the Ame and Catherine Vennema professor of chemical engineering and the Arthur F. Thurnau professor at the University of Michigan in Ann Arbor; he was the 2009 president of the American Institute of Chemical Engineers. He earned his BS degree from the University of Illinois and his MS and PhD degrees from the University of Colorado. He is also the author of the *Elements of Chemical Reaction Engineering*, which is one of the main textbooks for chemical engineering students.

Scott and his students are well known for their work on the application of chemical reaction engineering principles to the petroleum industry. They have published over 200 research articles in areas such as wax deposition/gelation kinetics in subsea pipelines, asphaltene flocculation/deposition kinetics, scale deposition and acidization of petroleum wells. In 1996, he was a recipient of the Warren K. Lewis award from the American Institute of Chemical Engineers for his contributions to chemical engineering education. He is also a recipient of 11 named lectureships.

1

Introduction

1.1 Background of Wax Deposition

Wax deposition is a critical operational challenge to the oil and gas industry. As early as 1928, wax deposition was reported as an issue that "presents many difficult problems while being produced, transported, and stored" (Reistle, 1932, page 7). Wax deposition problems occur in a wide range of locations in the petroleum production chain, including flow lines, surface equipment, and topside facilities, and downstream refineries. In some of the extreme cases, it can also occur in well tubings.

The waxy components of crude oils, also known as n-paraffins, represent a group of n-alkanes with carbon numbers that are usually greater than 20 (Lee, 2008). These components are normally dissolved in the oil at reservoir conditions where the temperature is relatively high. However, as the crude oil leaves the reservoir and travels toward processing facilities, its temperature can decrease substantially and potentially fall below the wax appearance temperature (WAT) (Berne-Allen & Work, 1938). When the waxy components can precipitate out of the oil and form solids, resulting in slurries in the oil flow that require higher pressure drop for transportation. More importantly, the precipitation of these components on the inner surface of the pipe wall can lead to the formation of wax deposits, which often occurs on the tubing, the pipelines, and the process equipment (Reistle, 1932).

In early- to mid-1990s, the problem of wax deposition usually occurred during petroleum production on land or onshore resources (Reistle, 1932). In 1969, it was reported that the cost for wax control in U.S. domestic production amounted annually to $4.5–$5 million (Bilderback & McDougall, 1969). Because of easy access and management for these resources, the problem of onshore wax deposition can be addressed by relatively simple methods, including the optimization of the operating conditions (pipeline size, pressure, etc.). Heating of the pipeline or mechanical removal of the wax deposit was used occasionally and was generally not as prohibitive. It is during the late twentieth century that the problem of wax deposition has become increasingly challenging, as the production of petroleum fluids shifted from onshore resources toward offshore reservoirs around the world. A schematic of this shift is shown in Figure 1.1 (Huang, Senra, Kapoor, & Fogler, 2011).

Taking the United States as an example, large offshore reservoirs that are mainly by the coastlines of Louisiana, Texas, California, and Alaska have quickly become one of the most crucial elements to the United States' strategic development of energy resources (*Economic analysis methodology for the 5-year OCS Oil and Gas Leasing Program for 2012–2017*, 2011). While 20 million bbl of oils were produced from offshore in the Gulf of Mexico in 1995, this number has risen to 1400 million bbl in 2007 (Bai & Bai, 2012). The offshore petroleum fluids are usually transported in long-distance pipelines, which range from tens to hundreds of kilometers before they eventually reach onshore processing facilities (Golczynski & Kempton, 2006). The oil typically comes out of the reservoir at a temperature around 160°F and is cooled significantly as it is transported through the pipes on the ocean floor, where the water temperature is around 40°F. This temperature difference between the oil in the pipeline and the surrounding water on the ocean floor (160°F to 40°F) becomes the driving force that causes the oil in the pipeline to cool down. As the oil temperature decreases, the waxy components can precipitate out of the oil and form deposits on the pipe wall. The problem of wax deposition in the subsea pipeline has caused a series of problems for the flow assurance industry, including increased pressure drop needed for oil transportation and potential blockage of the pipeline. An example of a plugged

FIGURE 1.1
A schematic of the change from onshore to offshore in petroleum production in the late twentieth century. (From Huang, Z. et al., *AIChE J. 57*, 841–851, 2011.)

pipeline due to wax deposition reported by Singh, Venkatesan, Fogler, and Nagarajan (2000) is shown in Figure 1.2.

The problem of wax deposition has become such a flow assurance concern that its severity must be assessed in the design of nearly every subsea development across the world, including the Gulf of Mexico (Kleinhans, Niesen, & Brown, 2000), the North Slope (Ashford, Blount, Marcou, & Ralph, 1990), the North Sea (Labes-Carrier, Rønningsen, Kolnes, & Leporcher, 2002; Rønningsen, 2012), North Africa (Barry, 1971), Northeast Asia (Bokin, Febrianti, Khabibullin, & Perez, 2010; Ding, Zhang, Li, Zhang, & Yang, 2006), Southern Asia (Agrawal, Khan, Surianarayanan, & Joshi, 1990; Suppiah et al., 2010), and South America (Garcia, 2001). The locations of these oil fields that have been reported to have concerns of wax deposition are highlighted in Figure 1.3.

Significant operational hazards due to wax deposition have been reported over the past few decades. The U.S. Minerals Management Service reported 51 occurrences of severe wax-related pipeline plugging in the Gulf of Mexico between the years 1992 and 2002 (Zhu, Walker, & Liang, 2008). One of the most severe cases was reported by Elf Aquitaine in which a removal of a wax-related pipeline blockage cost as much as $5 million. The remediation of this blockage resulted in a 40-day shutdown of the pipeline, which added an additional loss of $25 million of deferred revenue (Venkatesan, 2004). The arguably most notorious incidence might be from the Staffa Field, Block 3/8b, UK North Sea, in which the problem of wax deposition, after several unsuccessful attempts for remediation, eventually led to the abandonment of the field and its platform (Gluyas & Underhill, 2003), leading to an estimated loss of as much as $1 billion (Singh, 2000).

One of the main approaches to prevent wax deposition in subsea operations is pipeline insulation. However, this solution could greatly increase the

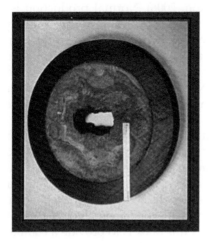

FIGURE 1.2
An example of a pipeline being plugged by wax deposits on the wall. (From Singh, P. et al., *AIChE J., 46*, 1059–1074, 2000.)

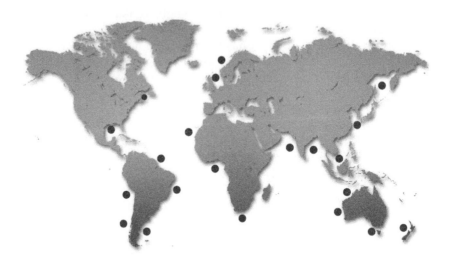

FIGURE 1.3
Areas reported to have wax deposition problems across the world, including the Gulf of Mexico (Kleinhans et al., 2000), the North Slope (Ashford et al., 1990), the North Sea (Labes-Carrier et al., 2002; Rønningsen, 2012), North Africa (Barry, 1971), Northeast Asia (Bokin et al., 2010; Ding et al., 2006), Southern Asia (Agrawal et al., 1990; Suppiah et al., 2010), and South America (Garcia, 2001).

production cost. For long-distance pipelines where a significant portion of the pipe is subjected to wax deposition risk, the most frequently used remediation method is called "pigging," which uses an inspection gauge with brushes or blades on its surface to scrape off the wax deposits on the wall (Golczynski & Kempton, 2006). Normal production is usually interrupted during the pigging operations, adding to the cost of production. The frequency of pigging can greatly influence the production cost. An estimate of deferred revenue based on a 29-km production pipeline and a production rate of 30,000 bbl/day with an oil price of $20/bbl at the time of the study is shown in Figure 1.4 (Niesen, 2002). It should be noted that the oil price nowadays has increased and thus, the production costs related to pigging will be much higher.

As we can see from the above analysis, it is extremely important to have a sufficient and rigorous understanding of the physics and chemistry of wax precipitation/deposition in the pipeline in order to develop economically viable prevention/mitigation strategies. The establishment of such an understanding can be achieved through a series of laboratory characterizations as well as predictive modeling that incorporates the fundamentals of thermodynamics and transport phenomena. The goal of this book is to provide a comprehensive introduction of the laboratory experiments and the thermodynamic/transport theories used for wax modeling, and to demonstrate how they can combine to deliver reliable solutions to address the problem of wax deposition in many cases.

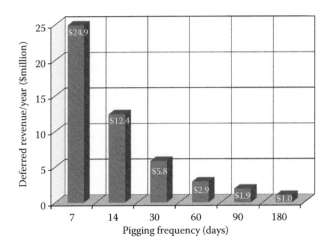

FIGURE 1.4
Cost of deferred production due to pigging estimated by Niesen (2002). (Niesen, V.: The real cost of subsea pigging. *E&P Mag.* 2002. pp. 97–98. Copyright Wiley-VCH Verlag GmbH & Co. KGaA. Reproduced with permission.)

1.2 Overview of Wax Testing, Modeling, and Management

A general overview of the general methodology used to develop experiments, predictions, and management strategies to address wax deposition is highlighted in Figure 1.5.

The first task to address in wax deposition is to understand how the waxy components (also known as n-paraffins) in a particular crude oil precipitate at different temperatures and pressures. It should be noted that the precipitation of these waxy components is mainly temperature-dependent, as pressure is often found to have an insignificant impact. To accomplish this task, thermodynamic modeling and experimental measurements are frequently used in combination to provide reliable information on wax precipitation characteristics, and these two aspects will be discussed in detail in Chapters 2 and 3. The output of this precipitation analysis is often visualized as a "precipitation curve of the wax," as shown in Figure 1.6.

In a wax precipitation curve, the highest temperature of a precipitation curve represents the WAT where the crystallization of waxy component starts to occur during cooling. At low temperatures, the precipitation curve usually extends to an asymptotic value, which represents the total wax content in oil. In this book, the experimental characterizations to determine the wax precipitation curve are highlighted in Chapter 2, while the theoretical modeling methods to predict the wax precipitation curve are introduced in Chapter 3.

A task of equal importance is to identify the hydrodynamic (i.e., pressure drop) and the radial and axial temperature profiles along the pipeline or

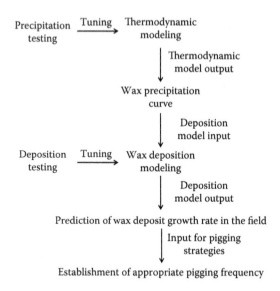

FIGURE 1.5
General overview of wax testing, modeling, and management procedures.

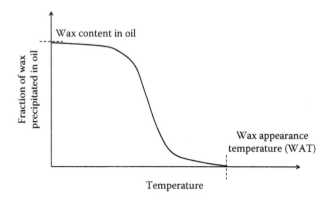

FIGURE 1.6
Schematic of a wax precipitation curve.

process equipment to reveal the location where wax deposition can potentially occur. This task can be completed using a wax deposition model that usually contains a flow simulator to determine the hydraulic and heat transfer characteristics in the pipe. The location where wax deposition starts to occur is usually where the inner wall temperature of the pipe drops to below the wax appearance temperature, as shown in Figure 1.7.

In addition to identifying the potential location of wax deposition in a pipeline, a wax deposition model can evaluate the severity of wax deposition by determining the growth rate of the wax deposits. To complete this task,

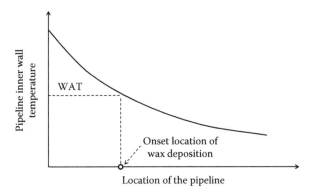

FIGURE 1.7
Schematic of the inner wall temperature determined by a wax deposition model to indicate the onset location for wax deposition.

fundamentals of transport phenomena are used by the wax deposition models, and they will be described in Chapter 4. The validity of the transport theories as well as the performance of the wax deposition models is usually evaluated through comparisons with flow loop wax deposition experiments. In the past few decades, many laboratory-scale or pilot-scale experiments have been carried out to study wax deposition. A summary of the laboratory equipment to study wax deposition is discussed in Chapter 5. In Chapter 6, examples of model validation using multiple experimental studies are presented, and the governing principles of the deposition phenomena are discussed. Through an in-depth theoretical analysis, it provides a revelation of the variables that have the most influence on the severity of wax deposition. Chapter 7 highlights the cases where wax deposition modeling has been applied to the field to identify the appropriate remediation strategies. Also included in Chapter 7 is a discussion of several paths forward to improve the current understanding of the wax deposition phenomenon in order to achieve more rigorous predictions in the field.

2

Experimental Characterization of Important Wax Thermodynamic Properties

The path to investigate the severity of wax deposition in field pipelines begins with a study of the key thermodynamic parameters involved in wax precipitation. This study provides the basic knowledge of wax precipitation that will enable us to make predictions of wax deposition. In this chapter, various experimental techniques will be used to characterize two key thermodynamic parameters involved in wax deposition from crude oils: the wax appearance temperature (WAT) and the wax precipitation curve (WPC). In Chapter 3, theoretical modeling of wax deposition will be introduced to extend our knowledge of wax deposition to pressures and temperatures similar to those found in field operations.

An understanding of wax thermodynamics will help us answer three very important three very important questions about the crude oil of interest:

- At what temperature will wax start to precipitate?
- How much wax will precipitate at a particular temperature?
- What are the properties of the precipitated wax?

Various experimental techniques have been developed to answer these questions. The principles, applications, and limitations of these experimental techniques will be discussed in the following.

2.1 Introduction

Let's begin with the attempt to answer the first question of wax deposition, "Do we have a problem for this oil field?" While a definitive answer to this question might require sophisticated fluid testing procedures, in many industrial practices, one can often get a general idea by just looking at a few parameters, for example, the wax content as well as the WAT. The values of these parameters can often be found in the fluid testing conducted on the stock tank oil sample of the production fluid. This testing is usually carried out during the early stage of the flow assurance analysis for a field. An example of the stock tank oil analysis is shown in Table 2.1.

TABLE 2.1

Example of Stock Tank Oil Analysis Results for Petroleum Fluid

Stock Tank Oil Analysis		
Drilling fluid content	<0.8	wt% STO
API gravity	32.1	at 60 °F
Wax content	4.5	wt%
Asphaltene content	1.8	wt%
Sulfur content	0.9	wt%
Wax appearance temp. (CPM)	95	°F
Pour point	30	°F
Total acid number	0.07	mg KOH/g

Intuitively, we would think that an oil with more than 10 wt% wax content is more likely to cause wax deposition problem than another oil with less than 2 wt% of wax, and that an oil with a WAT of 120°F (49°C) is likely to be more difficult to manage than another oil with a WAT of 50°F (10°C). Generally speaking, crude oils with more than 2% wax content and a WAT higher than the ocean floor temperature (39°F or 4°C) could have wax deposition risks. However, such thresholds only provide a very rough and often overly conservative estimation. One can possibly find cases that an oil with a WAT higher than 39°F does not exhibit wax deposition during production. In addition, two oils sharing similar wax content can have a great difference in their deposition tendency. More importantly, one might find that two characterizations on the same stock tank oil can show drastically different wax contents or WAT values. In this case, how do we appropriately interpret the test results and identify which one is more representative of the field?

In this chapter, we will go through a variety of testing methods in great detail in order to introduce a comprehensive understanding on the laboratory characterizations that would provide the data that are crucial to reliably determine if a petroleum field is prone to the risk of wax deposition. These characterizations focus on the two key thermodynamic parameters: the WAT and the WPC. In Chapter 3, theoretical modeling of wax deposition will be introduced as an alternative approach to study wax precipitation.

2.2 Determination of the WAT

As the temperature of the crude oil decreases, the waxy components start to precipitate and form solid crystals. The WAT usually refers to this onset temperature at which the precipitation of wax is observed. The WAT is often one of the first parameters used in the design of offshore drilling facilities because it indicates the possibility of observing wax deposition in oil transportation as well as the location of the onset for wax deposition.

Most of the WAT measurements utilize the change in the physical properties of the oil during the formation of solid wax crystals. In all techniques for WAT detection, wax precipitation is first induced by a controlled cooling of the oil sample. At a temperature below but close to the WAT, the "first few" wax crystals form, and the resulting change in physical properties can be captured by the appropriate instruments; the WAT temperature can thus be determined.

It should be noted that the definition of the "first few" wax crystals is rather loose. For example, using a microscope with 125× magnification, one can observe wax crystals with a size of ~1 μm (Rønningsen, Bjamdal, Hansen, & Pedersen, 1991). Typically, wax crystals can grow to a size of >10 μm. Therefore, the size of a ~1 μm crystal indicates that the wax might be at its early to middle stages of formation (Venkatesan, 2004). However, even the ~1 μm wax crystal is not necessarily the "very first" formed. Using more advanced observation techniques with higher size resolution, such as near-infrared (NIR) scattering, one can observe crystals with an even smaller size below ~55 nm (~1/200 of the size limit of microscopy) (Paso, Kallevik, & Sjöblom, 2009). Even with NIR, one cannot conclude that the wax crystal smaller than 55 nm in size is the very first crystal. Actually, there is no detection technique that can detect the exact onset of wax precipitation (the thermodynamic WAT), which lies on the solid–liquid phase boundary (Coutinho & Daridon, 2005). At the exact thermodynamic onset temperature of wax precipitation, the size of the wax crystal is infinitesimally small, and instrumental techniques require some amount of solid wax to be present in order for the change in physical properties to be detectable. Consequently, it is more accurate to say that the WAT determined by instruments represents the temperature at which a observable amount of wax has precipitated (instead of saying that the WAT is *the* onset of wax precipitation). Depending on the variation in the detection limits, the WAT determined by different techniques can vary as much as 20°C (Coutinho & Daridon, 2005). Due to the variations in the WAT determined by different techniques, tests using different techniques are usually performed and cross-compared to establish bounds on the WAT.

Although different techniques detect the presence of wax crystals based on different physical properties of the sample, the general strategy applied in each technique is similar, as summarized in Figure 2.1.

Optical techniques detect the presence of solid wax crystals based on the interaction between the crystalline material and light. Techniques in this category include visual inspection, cross-polarized microscopy (CPM), and Fourier transform infrared spectroscopy (FT-IR). These three techniques will be discussed in Sections 2.1.1 through 2.1.3. Rheological techniques such as viscometry can be used to detect the presence of suspended solid wax particles based on the change in bulk viscosity. Determination of WAT based on

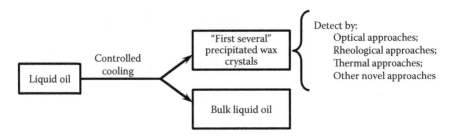

FIGURE 2.1
General strategy of WAT detection.

viscometry will be discussed in Section 2.1.4. Thermal techniques such as differential scanning calorimetry (DSC) identify the change in released heat due to crystallization of wax and will be discussed in Section 2.1.5. The comparison between the WAT measured by various techniques will be provided in Section 2.1.6. Other novel but not yet widely adopted techniques will be briefly introduced in Section 2.1.7.

2.2.1 Visual Inspection

Solid wax crystals scatter light, and thus the oil–wax suspension appears opaque. Using visual inspection, one can determine the WAT to be the temperature at which cloudiness of the oil can first be observed due to wax precipitation. Consequently, the WAT is also known as the cloud point. Standard testing methods, such as ASTM-D2500, are developed based on the visual inspection of wax solids by the naked eye (ASTM International, 2011). Figure 2.2 shows the experimental apparatus for the determination of the WAT using ASTM-D2500.

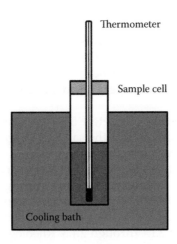

FIGURE 2.2
Apparatus for cloud point test using ASTM-D2500.

A sample of crude oil is placed in a test jar and then submerged in a cooling bath at a constant temperature. The decrease in the sample temperature is monitored by a thermometer. After the temperature of the sample is decreased by 1°C, the test jar is removed from the cooling bath and inspected for signs of solid precipitation. It should be noted that excessive temperature differences between the cooling bath and the sample can lead to supercooling of the sample and thus delay the formation of solid crystals (Bhat & Mehrotra, 2004; Coutinho & Daridon, 2005). Therefore, it might appear to make sense to utilize computer-programmed controllers to impose more carefully controlled slow cooling steps to alleviate the effects of supercooling on wax crystal formation. However, even with an infinitesimally slow cooling rate, the exact onset of wax precipitation is still experimentally inaccessible due to the aforementioned detection limits of the apparatus. Moreover, when the cooling rate approaches zero, measurement of the WAT is unreasonably long and tedious. Practically speaking, when the sample is cooled at a finite cooling rate ranging from 0.1°C/min to 10°C/min (Hansen, Larsen, Pedersen, Nielsen, & Rønningsen, 1991; Rønningsen et al., 1991). Within this practical range of cooling rates, supercooling is likely to occur, and thus the measured WAT will vary with the cooling rate.

Regardless of the WAT detection methods used, wax precipitation must be first induced by cooling the sample. Therefore, uncertainties in the WAT caused by supercooling can be considered as a universal limitation in WAT detection techniques. In general, the measured WAT decreases as the cooling rate increases because of the increasing extent of supercooling. Table 2.2 summarizes the effects of the cooling rate on the WAT when different techniques are applied.

It should be noted that the effect of the cooling rate on the WAT measured by ASTM is not listed in Table 2.2 because the cooling rate in the ASTM is not controlled.

The reliability of the WAT determined by ASTM-D2500 will vary for oil with different waxy components. For example, for a model n-C_{12} (oil) and n-C_{16} (wax) system that is transparent and colorless, the WAT measured by ASTM-D2500 correlates well within ~1°C difference of the values provided by DSC and viscometry (Bhat & Mehrotra, 2004). Crude oils are usually dark in color due to the light absorbance of the many other components such as aromatics. Therefore, the determination of the "first" haziness during cooling

TABLE 2.2

Summary of the Effect of Cooling Rate on the Variation in Measured WAT

Technique	Variation in Cooling Rates	Decrease in WAT	Source of Studies
CPM	0.1°C/min to 0.5°C/min	~3°C/min	Rønningsen et al. (1991)
DSC	1°C/min to 3°C/min	<1°C/min	Alghanduri, Elgarni, Daridon, and Coutinho (2010)
Viscometry	0.03°C/min to 0.2°C/min	<1.5°C/min	Rønningsen et al. (1991)

can be more difficult than in the case of model mixtures. The crude oil WAT determined from ASTM-D2500 seems to be consistently ~5°C lower than the values determined by DSC and viscometry (Claudy, Letoffe, Neff, & Damin, 1986; Kruka, Cadena, & Long, 1995). Moreover, ASTM-D2500 is not applicable when the color of the sample is darker than the ASTM-D1500 color of 3.5.

ASTM-D3117 was developed from the ASTM-D2500 based on the motivation to facilitate visual observation of haziness in dark samples. In ASTM-D3117, a smaller sample cell with an inner diameter (I.D.) of 0.8 in. is used in place of the ~1.2 in. I.D. sample cell used in ASTM-D2500. Using a thinner sample cell, the thickness of the liquid sample film is reduced, and the crude oil will appear less dark and thus allow better observation of the "first haziness." However, the improvement from ASTM-D2500 to ASTM-D3117 might still not be sufficient to determine the WAT of dark crude samples. Ijeomah et al. (2008) attempted to use ASTM-D3117 to measure the WAT of a synthetic crude and an Alaska North Slope (ANS) pipeline crude. In the study by Ijeomah et al., it was reported that ASTM-D3117 is only applicable to the GTL sample because it is transparent and light in color, while the determination of the WAT of the ANS crude is impossible due to its dark color.

As is presented in this section, ASTM-D2500 and ASTM-D3117 provide two rapid approaches to determine the WAT. However, the applicability of these two methods in the determination of the crude oil WAT is greatly limited due to the dark color of the crude oil samples. In addition, the uncontrolled cooling in these two testing methods is likely to cause delay in wax precipitation and underestimation of the WAT.

2.2.2 CPM Techniques

CPM has the following advantages over ASTM-D2500 and ASTM-D3117:

- It uses magnification lenses to extend the detection limits to smaller crystal sizes.
- It applies cross-polarized light to allow the detection of solids in dark oil samples.

The procedure to detect the WAT using CPM techniques is similar to that of visual inspection: A drop of liquid oil sample is placed on a glass slide and covered with a top glass to form a ~50 μm liquid film. The liquid film is then cooled at a cooling rate of usually 0.5 to 1°C/min, and the first sign of crystal formation is monitored by the microscope. Using a microscope, the object can be magnified by ~100-fold, and one can observe solids as small as ~0.5 to 1 μm. In addition, the contrast between the wax solid and the liquid is enhanced using cross-polarized light instead of unpolarized light. Under a cross-polarized microscope, the liquid oil appears black because cross-polarized light cannot pass through two perpendicular Nicol prisms, as shown in Figure 2.3a. However, crystalline materials such as wax crystals are optically anisotropic and therefore can rotate the plane of polarization of the polarized light, allowing a portion of the cross-polarized light to pass through the Nicol prisms, as shown in Figure 2.3b.

Consequently, wax crystals appear bright under a cross-polarized microscope as they form while the liquid remains black. The enhanced contrast between liquid and solid extends the applicability of the CPM methods to detecting the WAT of crude oil samples with a relatively dark color. Figure 2.4 shows a representative micrograph showing wax precipitating from a North Sea crude oil sample (Rønningsen et al., 1991). Due to these advantages, CPM has been widely used for the measurement of WAT of crude oils. Table 2.3 summarizes studies where the determination of WAT using CPM is reported.

It has been reported in multiple studies that CPM consistently gives WATs higher than those measured by other techniques, e.g., viscometry and DSC (Erickson et al., 1993; Monger-McClure et al., 1999; Rønningsen et al., 1991). In some cases, the WAT measured by CPM can be more than 10°C higher than the WAT measured by viscometry and DSC. While CPM is able to detect wax crystals when they reach ~1 μm in size, the heat effects and the change in bulk viscosity associated with such small amount of precipitation are usually undetectable by DSC or viscometry. It usually requires 0.3–0.4 wt% solid

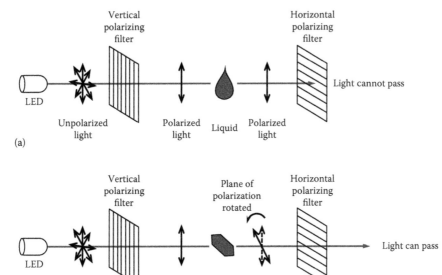

FIGURE 2.3
Interactions between the cross-polarized light and (a) homogeneous liquid and (b) crystalline material.

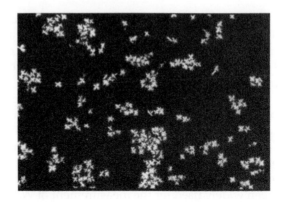

FIGURE 2.4
Photomicrograph of wax crystals under cross-polarized microscope; wax crystals appear as white flocs. (From Rønningsen, H. P. et al., *Energy & Fuels*, 5, 895–908, 1991.)

suspended in the liquid bulk in order for the change in viscosity to be detectable (Rønningsen et al., 1991).

It should be noted that in ASTM-D2500, ASTM-D3117, and the CPM method, detection of the WAT is often dependent on the operator's "judgment call" of "observing the first cloudiness" or "observing the first crystals." Consequently, the WAT determined by different observers can vary. Rønningsen et al. (1991) pointed out that the measurements made by one operator normally are repeatable within ~1°C, whereas the results obtained by different operators agree within 1.5°C–2°C. In order to reduce the subjectivity in the measurements, Kok et al. (1996) utilized a photomonitor to track the variation in the intensity of transmitted light through the sample, reducing the subjectivity in the WAT measurements due to different operators.

The CPM technique for the determination of the WAT overcomes the major drawbacks in ASTM-D2500 and ASTM-D3117 and has become the most applied technique in the determination of the WAT of crude oils. However, it should be noted that both ASTM-D2500/D3117 and CPM methods are based on the operator's visual detection of the first cloudiness. Therefore, measurements by different operators can vary, reducing the reliability of these two

TABLE 2.3

Summary of Experimental Studies on the Determination of WAT Using CPM Techniques

Studies	Sample Origin	Number of Samples
Rønningsen et al. (1991)	North Sea	17
Erickson, Niesen, and Brown (1993)	Gulf of Mexico	15
Monger-McClure, Tackett, and Merrill (1999)	Gulf of Mexico, Trinidad, Michigan, and Oklahoma	13
Cazaux, Barre, and Brucy (1998)	Undisclosed	2
Paso et al. (2009)	Undisclosed	2

methods. In Sections 2.1.3 through 2.1.5, WAT detection techniques based on measurable physical properties will be discussed. It is expected that the subjectivity in these techniques can be greatly reduced.

2.2.3 Fourier-Transform Infrared Spectroscopy

FT-IR techniques measure the WAT based on the distinct IR absorbance of solid wax compared with liquid oil. Consequently, the WAT measured by FT-IR techniques is not subject to the variations in the judgment calls by different operators as in the case with the ASTM D-2500, -D3117, and CPM techniques. The principles and applications of FT-IR in WAT determination will be discussed in this section.

Long-chain hydrocarbons with more than four consecutively connected methylene groups ($-CH_2-$), such as wax (n-alkanes), selectively absorb IR light with a wavenumber (often noted as v) of ~720 cm^{-1} to excite the rocking vibrational mode of the hydrocarbon chain, as shown in Figure 2.5 (Smith, 1999; Streitwieser & Heathcock, 1976). The absorption of the $v = 720$ cm^{-1} light results in a peak in a FT-IR absorption spectrum, as shown in Figure 2.6. The absorption by long-chain hydrocarbons is known as A_{720}.

In crystalline wax, the 720 cm^{-1} absorption band splits into two bands at 730 and 722 cm^{-1} due to the distinguishable in-phase covibration and out-of-phase covibration between the closely packed alkane chains in the crystal lattice (Smith, 1999). A typical IR absorption spectrum of crystalline wax is shown in Figure 2.7.

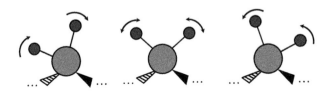

FIGURE 2.5
$-CH_2-$rocking vibrational modes in n-alkane chains.

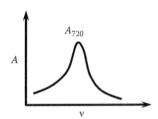

FIGURE 2.6
IR absorption spectrum of long-chain hydrocarbons with a characteristic absorbance at $v = 720$ cm^{-1}.

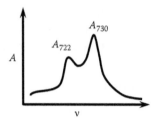

FIGURE 2.7
IR spectrum of crystalline long-chain hydrocarbons with the original 720 cm⁻¹ absorption split into two bands at 730 and 722 cm⁻¹.

The absorbance, A, cannot be used to correlate the amount of crystalline wax present because it varies with the shape of the IR absorption spectrum as the temperature changes (Roehner & Hanson, 2001).

Instead of the absorbance, the intensity, I, defined as the integrated area below the spectrum, as shown in Figure 2.8, is used to correlate with the amount of crystalline wax present in the sample.

The appearance of the wax crystals can alter the intensity of the liquid phase. At this point, the intensity represents a sum of the two components from both liquid and crystalline wax, as shown in Equation 2.1:

$$I_{total} = x_{liquid}I_{liquid} + (1 - x_{liquid})I_{crystalline} \qquad (2.1)$$

Although the exact reason is not yet completely understood, it is observed that the intensity of absorbance by solid alkanes, $I_{crystalline}$, is often ~50% greater than that of the liquid alkanes, I_{liquid} (Snyder, Hallmark, Strauss, & Maroncelli, 1986). Consequently, as wax crystallizes upon cooling, one observes an increase in the intensity, as shown in Figure 2.9.

The inflection point on the intensity vs. temperature curve is thus noted as the WAT. Table 2.4 summarizes several studies where the WAT of crude oils were measured using the FT-IR method.

In this section, the FT-IR technique for the determination of the WAT was discussed. FT-IR can be applied to measure the WAT of both the model

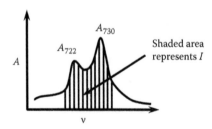

FIGURE 2.8
Definition of the intensity of IR absorbance as a quantitative variable to correlate to the amount of crystalline wax in oil.

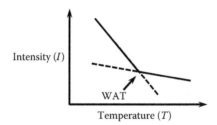

FIGURE 2.9
A drastic increase in the intensity as wax crystals form.

TABLE 2.4

Summary of Experimental Studies on the Determination of WAT Using
FT-IR Techniques

Literature	Sample Origin	Number of Samples
Roehner and Hanson (2001)	Gulf of Mexico, Utah-Grand County, and Alaska North Slope crude oil	3
Monger-McClure et al. (1999)	Gulf of Mexico, Trinidad, Michigan, and Oklahoma	13
Alcazar-Vara and Buenrostro-Gonzalez (2011)	Gulf of Mexico	3

mixture and crude oils that are dark in color. The sensitivity of the detection of the WAT by FT-IR can be comparable to the CPM method. However, it is sometimes difficult to determine the inflection point in the intensity–temperature curve and thus difficult to determine the WAT. The uncertainties in the determination of the inflection point prevail when the crude oil of interest has a relatively wide n-alkane distribution or a low total n-alkane content (Coutinho & Daridon, 2005).

2.2.4 Viscometry

Above the WAT, the crude oil behaves as a Newtonian fluid, and the viscosity of oil as a function of temperature can be described using the following Arrhenius-type equation:

$$\mu = Ae^{Ea/RT} \tag{2.2}$$

where E is typically in the range of 10–30 kJ/mol, and A is around $1–5 \times 10^{-3}$ Pa s (Rønningsen et al., 1991). As temperature is lowered below the WAT, the wax precipitates, and the precipitated solids remain suspended in the bulk liquid. The suspended wax particles change the flow properties of the crude oil. At temperatures slightly below the WAT, the oil usually remains as a Newtonian fluid, but the viscosity increases at a higher rate with decreasing temperature compared to crude oil above the WAT (Li & Zhang, 2003;

Yan & Luo, 1987). Determination of the WAT based on viscometry utilizes the change in the slope in the viscosity vs. temperature curve. In a typical experimental measurement of viscosity using a rheometer, the oil sample is cooled at a rate of 0.03°C–2°C/min (Rønningsen et al., 1991). The change of viscosity as a function of temperature is measured under a constant shear rate between 30 and 300 s^{-1} (Rønningsen et al., 1991). In terms of reproducibility, the viscosity in the Newtonian region may have an error bar within 0.5%–1%, while the viscosity in the non-Newtonian region can achieve an error bar within 10% (Rønningsen et al., 1991).

In the viscosity–temperature curve of a fluid, extrapolation from the Newtonian region is performed to determine the WAT, as shown in Figure 2.10. The fitting can be done as an exponential fitting of μ as a function T using a computer, shown in Figure 2.10a or a semilog paper with ln(μ) as a function of $1/T$, shown in Figure 2.10b. Table 2.5 summarizes some of the representative studies in which the WAT of the crude oils is determined based on viscosity measurements.

It should be noted that when the temperature is further lowered below the WAT, a substantial amount of wax can precipitate and remain suspended in

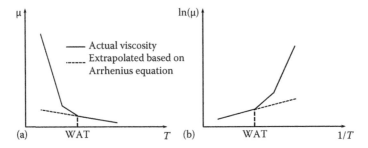

FIGURE 2.10

(a) Plot of a representative crude oil viscosity as a function of temperature with the dashed line showing the extrapolation from the Newtonian region of the curve. (b) Plot of ln(μ) as a function of $1/T$ with the dashed line showing extrapolation from the Newtonian region.

TABLE 2.5

Several Representative Studies Reporting the Measurement of WAT Based on Viscosity Measurements

Literature	Sample Origin	Number of Samples
Rønningsen et al. (1991)	North Sea	17
Kruka et al. (1995)	Middle East	1
Kok et al. (1996)	Undisclosed	15
Cazaux et al. (1998)	Undisclosed	2
Kok et al. (1999)	Undisclosed	8
Elsharkawy, Al-Sahhaf, and Fahim (2000)	Middle East	8
Alcazar-Vara and Buenrostro-Gonzalez (2011)	Gulf of Mexico	3
Erickson et al. (1993)	Gulf of Mexico	15

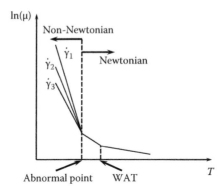

FIGURE 2.11
Change of crude oil viscosity during cooling processes. (From Li, H., & Zhang, J. *Fuel, 82,* 1387–1397, 2003.)

the crude oil, causing the shear-thinning characteristics of the crude oil, as shown in Figure 2.11.

In this section, the determination of the WAT based on viscometry was discussed. The sensitivity of the viscometry technique is lower than CPM and FT-IR. In most cases, an amount of 0.3–0.4 wt% solid material needs to have precipitated in order to cause a detectable deviation in the viscosity, whereas the wax crystals only need to grow to ~1 μm to be detectable by CPM, corresponding to ~0.1 wt% solid material in the liquid (Rønningsen et al., 1991). Therefore, the WAT determined based on viscosity measurements can be 0°C–10°C lower than the values determined by CPM (Erickson et al., 1993; Rønningsen et al., 1991). Depending on the composition of the precipitating n-alkane, the change in the slope of the viscosity–temperature curve near the WAT can be either sharp or gradual. If the precipitating n-alkane only covers a narrow range of carbon numbers, the change in the slope is likely to be sharp because n-alkanes precipitate in a narrow temperature range. Contrarily, if the precipitating n-alkane covers a wide range of carbon numbers, the change in the slope is likely to be gradual as the n-alkanes precipitate over a wider temperature range. Sometimes, the change in the slope can become difficult to identify when very little wax is precipitating, which will cause a significant underestimation of the WAT (Kok et al., 1996; Kruka et al., 1995).

2.2.5 Thermal Techniques—DSC

Wax precipitation is an exothermic process. Thermal techniques, e.g., DSC, capture the heat of crystallization released by the sample upon wax precipitation and thus detect the onset of wax precipitation. It should be noted that due to the thermal resistance of the sample and thus the finite rate of heat transfer, the actual temperature of the sample does not respond instantaneously to changes in the temperature. When the sample is cooled at a rate of −10°C/min, the temperature reading provided from the instrument can be ~5°C lower than the actual temperature of the sample (Hansen et al., 1991). Therefore, in order to obtain reliable values for the WAT, the temperature scale of the DSC instrument needs to be calibrated to account for the lag in the sample temperature compared with the temperature set point. The calibration can be conducted by measuring the melting point of known materials with high purity (e.g., indium's melting point at 156.60°C) using the same cooling procedure as that used for crude oil WAT measurements. Detailed operating procedures for DSC temperature calibration can be found in ASTM-E967-08 (ASTM International, 2008).

After the instrument is calibrated, an oil sample is placed in an ~40–75 μL stainless steel/aluminum capsule and cooled at a rate of ~0.5°C–10°C/min while the heat flow from the sample is recorded. A comprehensive procedure for the determination of the WAT using DSC can be found in ASTM-D4419-90 (ASTM International, 2010).

The procedures used in other studies resemble this standard testing method, while some case-specific parameters, such as the scanning rate and the amount of the sample, may vary from study to study. At the WAT, the heat of crystallization is released, resulting in an increase in heat flow. It should be noted that even without wax precipitation, the cooling of the oil sample requires heat to be removed from the sample. This heat removed from the sample will also be recorded by DSC (Hansen et al., 1991). Consequently, in order to isolate the heat effect due to wax precipitation from the heat effect due to temperature decrease, a baseline of the DSC thermogram needs to be defined. In a typical crude oil WAT characterization using DSC, one can define a baseline by connecting the starting point and the end point of a DSC exotherm, as shown in Figure 2.12 (Hansen et al., 1991).

The actual thermogram deviates from the baseline due to wax precipitation, as can be seen from Figure 2.12. The intersection of the baseline with the tangent of the leading peak edge at its inflection point is often defined as the

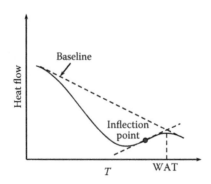

FIGURE 2.12
The definition of baseline and the WAT based on a typical DSC thermogram.

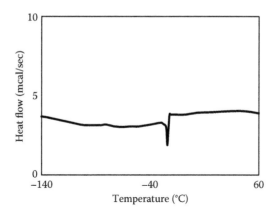

FIGURE 2.13
DSC thermogram of North Sea crude oil with narrow temperature range of wax precipitation. (From Hansen, A. B. et al., *Energy & Fuels, 5*, 914–923, 1991.)

WAT for this method (Standard Test Method for Measurement of Transition Temperatures of Petroleum Waxes by DSC, 2013). It should be noted that the shape of the thermogram greatly depends on the composition of the wax. A sharp peak on the thermogram can be observed when the precipitation of wax occurs in a narrow temperature range, as can be seen from Figure 2.13. The heat release peak can be gradual if the wax precipitation occurs in a wide temperature range. Due to the complex composition of the wax in the crude oil, its DSC thermogram can sometimes be irregular, as shown in Figure 2.14 (Hansen et al., 1991).

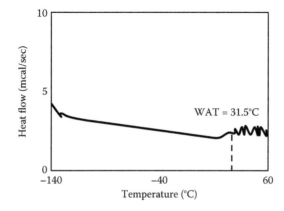

FIGURE 2.14
DSC thermogram of North Sea crude oil with irregular shape. (From Hansen, A. B. et al., *Energy & Fuels, 5*, 914–923, 1991.)

Due to these complications, the determination of the intersection between the baseline and the tangent of the peak front can become difficult (Hansen et al., 1991). In many cases, the WAT is determined as the temperature at which visible deviation of the thermogram from the baseline can be identified (Hansen et al., 1991). The standard deviation of the WAT determined from thermogram is ~2°C (Hansen et al., 1991).

Table 2.6 summarizes the experimental studies where the determination of the WAT by DSC is reported. Similar to other WAT measurements, DSC requires a sufficient amount of wax (~0.3–0.4 wt%) to crystallize in order for the heat effect to be detectable (Hansen et al., 1991). This value is generally greater than that for CPM, and thus the WAT determined from the DSC can be ~8°C lower than that determined by the CPM. However, the sensitivity of DSC in the determination of the WAT might be comparable with that determined by viscosity measurements, as both techniques require ~0.3–0.4 wt% of solid present at the WAT (Hansen et al., 1991). In addition to the amount of crystallization, the intensity of the DSC heat flow signal is also dependent on the rate of crystallization. The rate of crystallization decreases with a decreasing cooling rate. A low crystallization rate can make it difficult for the DSC to detect the heat of crystallization (Coutinho & Daridon, 2005). Therefore, a cooling rate that is too low could reduce the sensitivity of DSC, although it might alleviate the supercooling effect encountered at a fast cooling rate (Elsharkawy et al., 2000). Figure 2.15 shows a representative trend in DSC thermogram at different cooling rates.

As can be seen in Figure 2.15, the detected onset of heat release shifts toward the right as the cooling rate decreases, indicating that the impact of supercooling is reduced, and the measured WAT is closer to the thermodynamic WAT. However, the intensity of the heat release peak also decreases as the cooling rate is lowered. If the cooling rate is further lowered, one would expect the intensity of heat release peak to continue to decrease and eventually become overshadowed by the noise.

TABLE 2.6

Summary of Studies Reporting Measurement of WAT Using DSC

Literature	Sample Origin	Number of Samples
Hansen et al. (1991)	North Sea	17
Kok et al. (1996)	Undisclosed	15
Monger-McClure et al. (1999)	Gulf of Mexico	13
Kok et al. (1999)	Undisclosed	8
Elsharkawy et al. (2000)	Middle East	8
Alghanduri et al. (2010)	Libya	5
Alcazar-Vara and Buenrostro-Gonzalez (2011)	Gulf of Mexico	3
de Oliveira et al. (2012)	Brazil	2

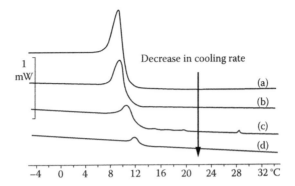

FIGURE 2.15
DSC thermogram obtained by cooling a crude oil sample at (a) 10°C/min, (b) 7°C/min, (c) 4°C/min, and (d) 1°C/min. (From Jiang, Z. et al., *Fuel, 80,* 367–371, 2001.)

The DSC method has two unique advantages over other detection methods. Firstly, DSC analysis can identify different temperature regions where wax precipitates (Kruka et al., 1995). Hammami and Mehrotra (1995) have demonstrated that using DSC, one can observe the precipitation of different n-alkanes occurring at different temperature ranges. Figure 2.16 shows a DSC thermogram of the $C_{50} + C_{44}$ binary mixture.

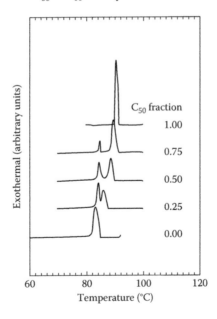

FIGURE 2.16
DSC thermogram of the $C_{50} + C_{44}$ binary mixture at 1°C/min cooling rate. The C_{50} fraction in the binary mixture varies from 1.00 to 0.00. (From Hammami, A., & Mehrotra, A. K., *Fluid Phase Equilib., 111,* 253–272, 1995.)

It is seen from Figure 2.16 that two distinct peaks can be identified from the thermogram. The peak in the higher temperature region is associated with the crystallization of C_{50}, and the other peak in the low temperature region is due to the crystallization of C_{44}.

Secondly, DSC can be adapted to determine the WAT of live crude oils that often contain a significant amount of light ends. The light ends in live crude often have a bubble point below atmospheric pressure. The effect of the light end composition on the crude oil WAT can be profound (Juyal, Cao, Yen, & Venkatesan, 2011). In this case, the WAT increases as the amount of dissolved light end decreases. In the case of a North Sea gas condensate, an increase as large as 10°C in the WAT is observed due to depressurization and the loss of light end (Daridon, Coutinho, & Montel, 2001). During a DSC test, one can maintain the dissolved gas in live crude by regulating the sample pressure over a range of 1–1000 bars using a specific high-pressure testing cell. The WAT of live crudes can thus be measured (Juyal et al., 2011; Vieira, Buchuid, & Lucas, 2010).

In this section, the determination of the WAT by DSC was discussed. The sensitivity of DSC in determining the WAT is comparable to that of viscometry but is less than that of CPM and FT-IR. The DSC method has two advantages over other techniques: it can detect different temperature ranges of wax precipitation and can be adapted to measure the WAT of live crude oils at elevated pressure.

2.2.6 Comparison between Different WAT Measurements

At this point, one might ask the question, "How do the WAT values by these four techniques compare?" Over the past two decades, the CPM method, the viscometry method, the DSC method, and the FT-IR method have been extensively compared in multiple comparative studies. Table 2.7 summarizes some of the most representative ones that were chosen to be highlighted in this book.

Due to the low detection limit of the CPM method, it is likely that CPM provides the most conservative measure of the WAT among the four methods. Figure 2.17 summarizes the statistics of 104 pairs of comparison of the WAT by DSC and WAT by viscosity and the WAT by CPM.

TABLE 2.7

Four Representative Reviews Comparing the WAT Values Obtained from Different Detection Techniques

Literature	CPM	DSC	Viscometry	FT-IR
Erickson et al. (1993)	×	×	×	
Kruka et al. (1995)	×	×	×	
Monger-McClure et al. (1999)	×	×		×
Coutinho and Daridon (2005)	×	×	×	×

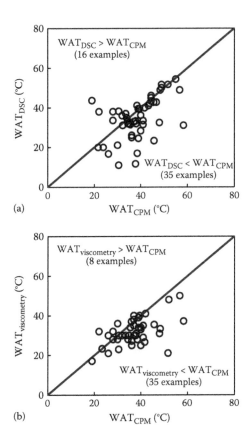

FIGURE 2.17
Comparison between the WAT measured by different methods: (a) DSC and CPM; (b) viscometry and CPM.

It is seen that, statistically, the WAT determined by CPM is higher than the WAT determined either by DSC or by viscometry, indicating that CPM provides the most conservative value (the highest estimation) for the WAT in most cases.

Different from the DSC and viscometry measurements, which generally provide lower WATs than CPM, FT-IR provides WAT values that are comparable with those measured by CPM, as shown in Figure 2.18 (Monger-McClure et al., 1999).

Meanwhile, the viscometry and DSC method usually provide comparable WAT values, as can be seen in Figure 2.19. The number of cases where the WAT (DSC) is greater than the WAT (viscometry) is similar to the number of cases where the WAT (DSC) is smaller than the WAT (viscometry), indicating that both techniques often share similar sensitivity.

In summary, for the WATs determined by the four techniques, one might wonder which is the "correct" WAT. Actually, none of these four WAT values

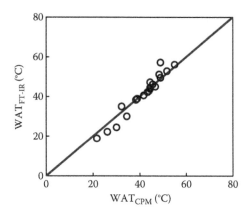

FIGURE 2.18
Comparison between the WAT obtained from FT-IR and the WAT obtained from CPM.

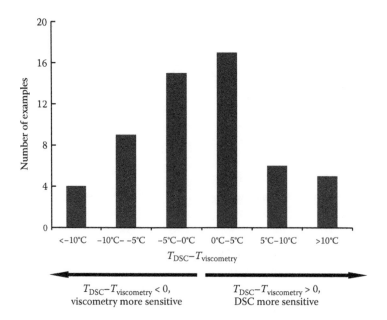

FIGURE 2.19
Comparisons between the WAT measured by the viscometry method and the DSC method.

is correct because none of them represents the exact onset of wax precipitation. The WATs provided by CPM and FT-IR are the closest to the exact onset at the thermodynamic WAT, whereas the WATs provided by DSC and viscosity are usually further below the thermodynamic WAT. The WATs by DSC and viscometry thus provide an estimation of temperature below which severe wax precipitation could occur.

2.2.7 Other Techniques That Are Under Development

In order to enhance the sensitivity of DSC, Jiang, Hutchinson, and Imrie (2001) explored the feasibility of using temperature-modulated DSC (TMDSC) to measure the WAT. Compared with regular DSC, in TMDSC, a sinusoidal temperature modulation is superimposed on a constant cooling rate. The temperature at different times and the cooling rate in a TMDSC can thus be described by Equation 2.3:

$$T = T_0 - \beta t + A_T \sin(\omega t)$$

$$\frac{dT}{dt} = -\beta + \omega A_T \cos(\omega t) \tag{2.3}$$

The term $-\beta$ represents the constant cooling rate, and the term $\omega A_T \cos(\omega t)$ represents the additional oscillation in the cooling rate imposed by TMDSC. In theory, the additional oscillation in the cooling rate enhances the rate of crystallization near the WAT. Unfortunately, the results reported by Jiang et al. have not yet demonstrated a significant improvement over conventional DSC (Coutinho & Daridon, 2005).

Other techniques to measure the WAT include NIR scattering, x-ray CT, densitometry, and filter plugging. Details regarding the principles and detection limits of these techniques are summarized in Table 2.8. Sample studies applying these techniques for WAT detection are also provided in Table 2.8.

TABLE 2.8

Summary of Principles, Detection Limits, and Sample Studies of NIR Scattering, X-Ray CT, Densitometry, and Filter-Plugging for WAT Detection

Technique	Principle	Detection Limit	Sample Study
NIR scattering	Solid wax crystals can scatter NIR light.	Theoretically more sensitive than CPM and can detect wax crystals with size ~55 nm	Paso et al. (2009)
X-ray CT	A solid wax crystal has a higher density than the liquid bulk.	Comparable to ASTM but can detect the WAT for dark crude oils, which cannot be analyzed using ASTM	Karacan, Demiral, and Kok (2000)
Densitometry		Comparable to DSC and viscometry	Alcazar-Vara and Buenrostro-Gonzalez (2011)
Filter-plugging	Solid wax crystals larger than ~5 µm in size clog a filter with pore size ~5 µm, resulting in a detectable increase in the differential pressure.	Comparable to CPM	Monger-McClure et al. (1999)

2.3 Determination of the WPC

In Section 2.1, it was shown that the detection of WAT depends on the change in physical properties caused by the presence of solid wax. In fact, some of these changes in physical properties are directly proportional to or can be well correlated with the actual amount of the solid wax. Therefore, in addition to the measurement of the WAT, the changes in these physical properties provide information regarding the amount of precipitated wax at any temperature below the WAT, which is a very important piece of information also known as the WPC, as shown in Figure 2.20.

The wax precipitation curve will be noted as WPC in this book. Table 2.9 summarizes the properties correlated to the amount of wax precipitation and the corresponding experimental techniques that directly measure these properties.

2.3.1 Differential Scanning Calorimetry

The WPC is most often measured by DSC in industrial practice. Figure 2.21 shows a schematic of a DSC thermogram of waxy oil during cooling.

As previously discussed in Section 2.1.5, in the DSC thermogram, the heat release curve deviates from its baseline due to wax precipitation. The cumulative heat released from the onset of precipitation to a certain temperature can be obtained by integrating the area enclosed by the thermogram and the baseline, as shown in Figure 2.21. This cumulative heat released ($q_{WAT \to T}$) is related to the amount of wax that has precipitated from a temperature above the WAT to a particular temperature, w_{wax}, as shown in Equation 2.4:

$$w_{wax,WAT \to T} = \frac{q_{WAT \to T}}{\Delta H_{crystallization}} \tag{2.4}$$

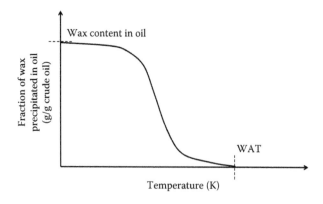

FIGURE 2.20
A representative example of the WPC of a crude oil.

TABLE 2.9

Summary of Different Experimental Techniques to Characterize the Wax Precipitation Curve

Technique	Property Directly Measured	Relation to the Amount of Precipitated Solid Wax	Examples
DSC	Heat flow due to precipitation	The heat flow released during precipitation is directly proportional to the amount of wax precipitated.	Hansen et al. (1991); Elsharkawy et al. (2000); Martos et al. (2008, 2010); Coto, Martos, Espada, Robustillo, Peña et al. (2011); Coto, Martos, Espada, Robustillo, Merino-García et al. (2011)
Nuclear magnetic resonance (NMR)	Resonance peak intensity	The peak intensity is directly proportional to the number of hydrogen atoms in the solid phase.	Pedersen, Hansen, Larsen, Nielsen, and Rønningsen (1991)
FT-IR	Intensity of FT-IR absorbance	The intensity of the characteristic FT-IR absorbance is directly proportional to the amount of solid wax.	Alcazar-Vara and Buenrostro-Gonzalez (2011); Roehner and Hanson (2001)

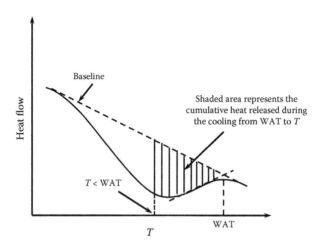

FIGURE 2.21
Representative DSC thermogram of waxy crude oil. The cumulative heat released due to wax precipitation can be determined by integrating the area enclosed by the baseline and the thermogram.

Knowing the heat of crystallization, $\Delta H_{crystallization}$, one can calculate the amount of solid wax that has precipitated when cooled from WAT to T according to Equation 2.4. It should be noted that determination of an appropriate value for $\Delta H_{crystallization}$ to be used in Equation 2.4 is a nontrivial matter. Depending on the origin of the wax, its enthalpy of crystallization can vary from ~100 to 300 J/g. Moreover, wax precipitated at different temperatures can have different chain lengths and thus different enthalpies of crystallization.

In absence of additional knowledge, the following assumptions regarding the enthalpy of crystallization are usually made to calculate the WPC:

- The enthalpy of crystallization of waxes is a constant (not changing with respect to temperature).

- The enthalpy of crystallization has a value of 200 J/g.

Both of these assumptions might not be valid and thus introduce uncertainties in the precipitation curve constructed from a DSC trace. The accuracy of the precipitation curve from DSC can be improved if one also knows the total wax content in the oil. Knowing the total wax content in the oil, one can scale the DSC precipitation curve until it matches the measured total wax content in the low temperature range. The total wax content can usually be measured using well-established protocols, such as UOP 46-64 or UOP 46-85, as well as other more advanced techniques such as the NMR.

It should be noted that scaling of the entire curve using the total wax content still assumes that the heats of crystallization for waxes precipitated at different temperatures are equal. This assumption might not be valid if the carbon number distribution of the wax spans a wide range. The carbon number of precipitated wax can vary with temperature. Thus, the heat of crystallization that should be used to calculate the amount of solid will vary with temperature. In order to determine what value of heat crystallization should be used to calculate the amount precipitated at different temperatures, knowledge of the composition of wax precipitated at different temperatures is required. This information can be predicted by a thermodynamic model. A method of improving the accuracy of the DSC curve using a predictive thermodynamic model will be discussed in Chapter 3.

Representative studies that characterized the WPC are summarized in Table 2.10.

TABLE 2.10

Summary of Experimental Studies on WPC Characterization with DSC

Literature	Characteristics of the Example(s)
Hansen et al. (1991)	1 North Sea crude oil (Oil 3 in the total 17 oils reported)
Juyal et al. (2011)	Synthetic oil with a light volatile crude oil and a gas blend prepared to match the flash gas composition from live crude
Martos et al. (2010)	3 Brazilian dead crude oils
Martos et al. (2008)	2 Brazilian dead crude oils

2.3.2 Characterization of WPC Using NMR

The purpose of this section is not to elaborate on the principles of NMR from a spectroscopist's point of view, as the basics of NMR can be found available in instrumental analysis and organic chemistry textbooks. This section will focus only on the NMR principles relevant to the determination of WPC.

When an external magnetic field, B_0, is applied to a chemical species, proton nuclear spins align in the same or opposite direction with the external field. The energy of the spins in the same direction with B_0 (parallel spin) is slightly lower than the energy of those in the opposite direction with B_0 (antiparallel spin). Therefore, the spin state in the same direction with B_0 is slightly more populated than the other spin state, resulting in a net magnetization in the same direction with B_0. When perturbed by a second external field B_1 in perpendicular to B_0, the net magnetization precedes with respect to B_0 at a unique frequency, ν. The precession of a nucleus generates a detectable alternating current (AC). The intensity of the AC decays exponentially as the nuclear magnetic spins "relax" from the perturbation by the external field B_1. The determination of WPC by NMR is based on the difference in the relaxation behavior of protons in the solid and liquid states. After a proton is excited by a radio frequency pulse, it relaxes to its equilibrium state by emitting the excess energy with respect to equilibrium state to its surroundings, resulting in decay in the NMR intensity (free induction decay [FID]). The protons in the solid state have shorter relaxation times. The intensity due to solid-state proton thus quickly decays to a negligible level. Compared with solid-state protons, the liquid-state protons have longer relaxation times, and therefore, their FID has a relatively long "tail."

In a suspension of solid wax in oil, both proton relaxations in the solid and liquid phases contribute to the FID, and thus, the FID of the entire sample is a combination of both solid-phase FID and liquid-phase FID. As can be seen in a typical FID of a mixture of solid wax suspended in oil, as shown in Figure 2.22, the fast FID of the solid-phase proton is observed right after the pulse excitation. After the solid-phase proton signal decays to a negligible level, a long tail due to the relatively slower FID of the liquid-phase protons can be observed.

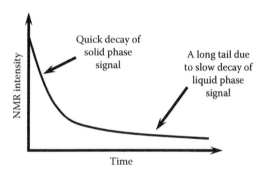

FIGURE 2.22
Representative NMR FID of a mixture with solid wax suspended in liquid oil.

In a representative NMR experiment for the determination of WPC, the NMR intensity of the same sample is measured at two times: (1) immediately after the radio frequency pulse ($t = t_1$) and (2) after the solid-state FID approaches zero ($t = t_2$). As the intensity due to the solid-phase proton quickly decays to a negligible level, the second NMR intensity thus only contains contribution from the liquid phase, while the first intensity contains both solid and liquid contributions. The two intensities can be used to back-calculate to the solid fraction in the sample based on the following mathematical derivation.

Let $I(T, t_1)$ and $I(T, t_2)$ be the two total intensities at t_1 and t_2, respectively. $I(T, t_1)$ and $I(T, t_2)$ can be expressed as linear combinations of the solid-state component and the liquid-state component, as shown in Equation 2.5:

$$t_1: \quad I(s, t_1) + I(l, t_1) = I(T, t_1)$$

$$t_2: \quad I(s, t_1)\exp\left(-\frac{t_2}{\tau_s}\right) + I(l, t_1)\exp\left(-\frac{t_2}{\tau_l}\right) = I(T, t_2) \tag{2.5}$$

As the intensities of the NMR signals, $I(s, t_1)$, $I(l, t_1)$, $I(s, t_2)$, and $I(l, t_2)$ are proportional to the amount of solid and liquid, one can determine the relative amount of solid and liquid by solving Equation 2.5 simultaneously.

Pedersen, Hansen et al. (1991) used the NMR technique to measure the WPC of 17 crude oils. Detailed procedures to perform this NMR measurement can be found in their published study.

In this section, the measurement of WPC by the NMR is discussed. Using the NMR technique, measuring the solid content at a particular temperature requires ~10 s. NMR measurements are expected to have a ~0.2% absolute standard deviation in the measured solid wt% of replicates of samples.

2.3.3 Characterization of WPC Using FT-IR

As discussed in Section 2.1.3, a long-chain alkane absorbs IR light at ~720 cm^{-1}. In a mixture with both solid and liquid alkanes, the intensity of IR absorbance is a linear combination of the contribution from the solid- and liquid-phase alkanes, as shown in Equation 2.6:

$$I_{\text{total}} = x_{\text{liquid,amorphous}} I_{\text{liquid,amorphous}} + (1 - x_{\text{liquid,amorphous}}) I_{\text{crystalline}} \tag{2.6}$$

The intensity of absorbance by the solid alkane, $I_{\text{crystalline}}$, is ~50% greater than that by the liquid or amorphous alkane, $I_{\text{liquid,amorphous}}$ (Snyder et al., 1986). Consequently, as crystalline wax forms upon cooling, one observes a significant increase in the intensity due to the contribution from solid wax, as previously shown in Figure 2.9 and as shown in Figure 2.23.

As shown in Figure 2.23, above the WAT of approximately 33°C, the intensity is solely due to the absorbance of the liquid phase. Thus, a linear

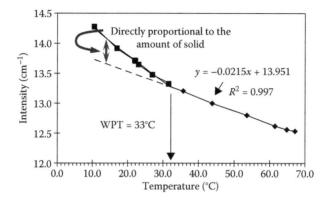

FIGURE 2.23
Intensity of FT-IR absorbance of a Gulf of Mexico crude oil as a function of temperature. (From Roehner, R. M., & Hanson, F. V., *Energy & Fuels*, 15, 756–763, 2001.)

extrapolation of the intensity from above the WAT to below the WAT, as denoted by a dashed line, provides an estimation of the liquid-phase contribution at temperatures below the WAT. The difference between the measured intensity and the extrapolation is directly proportional to the amount of solid present in the mixture. This difference can thus be used to back-calculate the amount of solid precipitated. Alcazar-Vara and Buenrostro-Gonzalez (2011) and Roehner and Hanson (2001) provide two representative studies regarding the application of FT-IR on WPC measurements.

Three techniques for the determination of the WPC, NMR, FT-IR, and DSC have been discussed so far. At this point, the readers might be interested in seeing a comparison between the WPC determined from different techniques. Unfortunately, studies reporting the application of these techniques are conducted individually with different crude oils, and there does not exist a comprehensive study using a single crude oil in order to make comparison of the WPC measured by these three techniques.

2.3.4 Separation-Based Methods for WPC Determination

In Sections 2.2.1 through 2.2.3, WPC determination was discussed based on the spectroscopic difference between the solid and liquid phase (FT-IR and NMR) and on the heat effect during phase transition (DSC). The theories behind these experimental techniques are somewhat abstract and may be difficult to comprehend without a lot of time and effort. One might wonder if it is at all necessary to rely on sophisticated instruments such as FT-IR, NMR, and DSC to characterize such a seemingly simple phase transition process of dissolved wax precipitating from the liquid bulk. Moreover, one might be tempted to cool the crude oil to various temperatures below the WAT and determine the WPC based on the weight of the solid cake precipitated at each temperature. However, the WPC determined in this way is not reliable because of the following reason: *Precipitation of wax molecules causes wax crystals to grow into an interlocking network that can entrap liquid oil.* Therefore, the weight of the "solid" cake actually represents the total weight of the "true" precipitated solid and the entrapped liquid.

Being aware of the issue that liquid oil can be trapped in the solid cake, one can improve the aforementioned procedures by either performing further removal of the entrapped oil or additional characterization of the precipitated solid phase to quantify its "true" solid content. In combination with proper removal of the entrapped oil or characterization of the solid cake, the proposed procedure based on separation of the solid and liquid phases can be used to obtain WPCs with acceptable accuracy. Table 2.11 summarizes representative studies where the oil precipitation curve is obtained based on separation of the solid cake from the liquid bulk.

TABLE 2.11

Summary of Representative Studies on WPC Characterization Using Separation-Based Techniques

Literature	Separation Techniques	Further Removal of Entrapped Liquid by	Characterization of the Solid Phase
Burger, Perkins, and Striegler (1981)	Centrifugation	Washing by toluene	Not performed
Coto, Martos, Peña, Espada, and Robustillo (2008)	Filtration	Washing by acetone	Not performed
Martos et al. (2008)	Filtration	Washing by acetone	Characterize the composition by HTGC, and calculate C_{15+} content to be the solid content
Roehner and Hanson (2001)	Centrifugation	Not performed	Characterize the composition of the solid by HTGC, and define the precipitated wax as the components with an n-alkane-to-nonalkane ratio in excess of the same component in oil
Dauphin, Daridon, Coutinho, Baylère, and Potin-Gautier (1999) Pauly, Dauphin, and Daridon (1998) Pauly, Daridon, and Coutinho (2004)	Filtration	Not performed	Characterize the composition of the solid by HTGC, and apply a mass balance to calculate the true solid content
Han, Huang, Senra, Hoffmann, and Fogler (2010)	Centrifugation	Not performed	Characterize the composition of the solid by HTGC, and apply a mass balance to calculate the true solid content

Attempts have been made to remove the entrapped oil by washing the solid phase with toluene or acetone (Burger et al., 1981; Coto et al., 2008). However, one cannot guarantee that washing by toluene or acetone completely removes the entrapped oil. Therefore, the WPC obtained after washing might still be an overestimation compared to the "true" WPC. Figure 2.24 shows a comparison of the WPC measured using fractional precipitation followed by acetone wash and the WPC of the sample crude measured by DSC.

It can be seen that the amount of solid precipitation measured by the fractional precipitation method is significantly higher than the values from the DSC method, indicating that acetone wash cannot completely remove the entrapped oil.

In order to provide a more accurate WPC, Martos et al. (2008) further characterized the composition of the solid cake using HTGC after washing with acetone. n-Paraffins with carbon numbers less than 15 have melting points lower than the room temperature. As a result, C_{15-} paraffins usually do not precipitate nor contribute to wax deposition. Martos et al. calculated the C_{15+} content of the solid cake based on the HTGC and used the calculated C_{15+} content to represent the "true" solid content of the solid cake. Martos et al.'s approach assumes that an n-paraffin with a carbon number less than 15 only exists in the entrapped liquid phase, while C_{15+} paraffins exist in the solid phase. Assuming that C_{15} is the lightest paraffin in the solid phase is purely empirical. Roehner and Hanson (2001) proposed a more rigorous method of determining the lightest n-paraffin component in the solid phase. They determined whether or not an n-paraffin component is in the solid phase by comparing its mass fraction in the solid cake with its mass fraction in the original oil. The precipitating paraffin components are enriched in the solid cake and thus have a higher mass fraction in the solid cake than in the crude

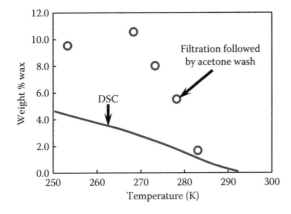

FIGURE 2.24
Comparison between the WPC determined by DSC and filtration followed by acetone wash. (From Martos, C. et al., *Energy & Fuels*, 22, 708–714, 2008.)

oil. Therefore, Roehner and Hanson count the amount of C_{i+} as the "true" solid content, where i is the lowest carbon number component whose mass fraction in the cake exceeds its mass fraction in the original oil. Both Martos et al. and Roehner and Hanson assume that a particular carbon number component exists only in the solid or liquid phase. Light components in the entrapped oil tend to exist in the liquid phase, whereas heavy components tend to exist in the solid phase. However, even heavy n-paraffins in the solid cake can exist in both liquid and solid phase: Heavy n-paraffins in the solid cake can remain dissolved in the entrapped oil. The heavy paraffins dissolved in the entrapped oil should not be counted as contributing to the solid phase. Therefore, the methods of Martos et al. and Roehner and Hanson might still overestimate the WPC. In order to account for the amount of dissolved heavy paraffin in the entrapped liquid phase, Dauphin et al. (1999), Pauly et al. (1998, 2004), and Han et al. (2010) proposed a more mathematically involved method where they solved a set of mass balance equations for the solid content of the solid cake.

In this section, separation-based techniques for the determination of the WPC are discussed. Without proper characterization of the solid content of the cake, these techniques tend to give overestimations of the WPC due to entrapment of oil in the solid cake. With proper characterization of the cake solid content, separation-based methods provide WPCs that are comparable with the precipitation curves measured by DSC, FT-IR, and NMR.

2.4 Experimental Techniques for the Characterization of Precipitated Wax

In addition to the WAT and the WPC, the mechanical properties of the wax deposit, such as the yield stress, provides key information regarding the easiness of the remediation of wax deposit. The yield stress of a particular wax deposit is dependent on the molecular structure and composition and of the wax responsible for the formation of this deposit. For example, according to the study by Petitjean et al. (2008), the yield stress of wax deposit depends on the molecular structure of wax. Wax deposits formed from branched and cyclic alkanes (also known as microcrystalline wax) are usually softer than deposits formed from the straight-chain alkanes, which are also known as macrocrystalline wax. Consequently, wax deposits formed by microcrystalline wax are easier to remediate when compared to deposits formed by macrocrystalline wax with the same carbon number. With information from structural characterization of the precipitated wax, one can provide recommendations regarding the mechanical properties and the easiness of deposit remediation. Table 2.12 summarizes the analytical methods that characterize the molecular structure of precipitated wax. In Table 2.12, the expected

TABLE 2.12

Summary of Experimental Techniques for the Characterization of the Composition and Structure of Wax Deposit

Analytical Methods	Measurable	Expected Experimental Outcome in the Case of Deposit Being Rich in Macrocrystalline Wax	Representative Studies
DSC	Heat of crystallization and melting point	High heat of fusion and narrow melting temperature range	Alghanduri et al. (2010)
XRD	Diffraction peaks	Can observe sharp diffraction peaks	Alghanduri et al. (2010)
^{1}HNMR	Numbers of methyl groups ($-CH_3$) and methylene groups ($-CH_2-$)	A low methyl-to-methylene ratio due to fewer branches	Musser, Kilpatrick, and Carolina (1998)
^{13}CNMR	Numbers of primary carbon ($-CH_3$), secondary carbon ($-CH_2-$), and tertiary carbon ($-CH-$) Number of carbon atoms on aromatic rings and cyclic rings	~70% of the carbon atoms exist in the form of primary ($-CH_3$) and secondary carbon ($-CH_2-$). Low content of tertiary carbon and carbon on rings	Musser et al. (1998); Alghanduri et al. (2010)
FT-IR	Intensity of absorption at ~720 cm^{-1} associated with long straight-chain methylene and the content of straight methylene	>60% straight methylene	Musser et al. (1998)
Elemental analysis	The ratio of the number of hydrogen atoms to the number of carbon atoms (H/C)	High H/C at ~2	Musser et al. (1998); Alghanduri et al. (2010)
Mass spectroscopy	Mass distribution of wax	MS peaks cover a range of 350–600 a.u., and two neighboring peaks are 14 a.u. (the molar mass of one methylene unit) apart	Musser et al. (1998)

experimental outcome for the deposit being rich in macrocrystalline wax is also summarized.

It should be noted that before applying the techniques listed in Table 2.12 to characterize precipitated wax, the precipitated wax should be purified by advanced separation techniques, such as column separation, in order to

remove entrapped oil. Otherwise, averaged properties of both the precipitated wax and the entrapped oil will be obtained instead of the properties of
the precipitated wax that will be obtained from characterization.

In addition, the yield stress of the deposit is also dependent on the composition of the wax in the deposit. Bai and Zhang (2013a) reported that the
yield stress of wax deposit varies as a function of the wax carbon number
distribution. Characterization of the composition of wax deposit can usually
be achieved by HTGC.

2.5 Summary

In this chapter, experimental techniques to characterize the following key
wax precipitation information were discussed:

- The WAT
- The WPC

The WAT provides information regarding the onset location of wax deposition in a transport pipeline. The WPC characterizes the solubility limit of
wax in crude oil and thus affects the concentration driving force for wax
deposition. The WAT and the WPC are both extremely important input
parameters required for wax deposition modeling. These experimental characterizations are also commonly used as references for benchmarking wax
thermodynamic models. The characterization of the composition and of the
molecular structure of the precipitated wax was also briefly introduced, as
this characterization provides a piece of information regarding the mechanical properties of the deposit.

It should be noted that the characterization of wax precipitation information is not always possible and becomes increasingly difficult and sometimes
virtually impossible in actual field operations. When experimental characterization of wax precipitation is difficult, modeling strategies are usually
applied to predict the wax precipitation characteristics. In Chapter 3, thermodynamic modeling of the wax precipitation characteristics will be discussed.

3

Thermodynamic Modeling
of Wax Precipitation

3.1 Introduction

In Chapter 2, we discussed different experimental approaches to characterize wax precipitation. The information of wax precipitation is not only indispensable to identify if wax deposition would be a problem for a field but also
important to quantify how severe the problem could be.

However, experiments are known to have their own disadvantages. First,
cost or time can be an issue. For example, the characterization of wax precipitation of a live crude oil that is rich in light ends would need to be performed in pressurized cells, which are often not readily accessible and the
experiments are expensive to perform. More importantly, reliability of the
results can be called into question. For example, for crude oils with low wax
content, the uncertainties in the measured WAT and WPC become large and
the results can be misleading. Although not quite frequently, sample mis-
handling or inappropriate design of experimental conditions can, in fact,
occur. Therefore, if we only rely on laboratory measurements, we might find
ourselves facing two results from similar testing methods showing quite different values. In this case, how would we gauge the validity of each result
without spending significant time and cost on a third test, which could also
be prone to uncertainties and errors? Sometimes experimental methods are
just not possible with today's technology: how can we conveniently take a
sample of wax deposit from an offshore subsea pipeline in the field to determine its composition, which could be an important piece of information for
one to determine the yield stress of the deposit and thus to optimize the pigging frequency?

Consequently, we need theoretical modeling as an alternative to provide
us answers that are not prone to the errors and uncertainties associated with
experimental measurements. The results of experimental data, when benchmarked with theoretical modeling, often provide the most reliable answers
for one to evaluate wax deposition concern for a field. In this chapter, we will

discuss the thermodynamic modeling of wax precipitation and demonstrate how it can help us better interpret the experimental characterizations.

Theoretical studies of wax precipitation started in the late 1980s (Won 1986). After a few decades, the thermodynamic theories of wax precipitation have gradually matured, and several commercial software packages have been made available for thermodynamic modeling of wax during common flow assurance practices. However, these software packages are still viewed as "black boxes" in many flow assurance specialists' perspective. In this chapter, we will first dissect the thermodynamic theories for wax precipitation (Sections 3.2–3.4). Understanding these theories can help us better understand the comparison of different commercial wax thermodynamic models, which will be discussed in detail in Section 3.5. We will finish the chapter with Section 3.6 that highlights several novel applications of wax thermodynamic modeling.

3.2 Fundamental Thermodynamics of Wax Precipitation

Modeling wax precipitation is not an easy task because there is not a pure substance with the name "wax" and another with the name "oil." Instead, wax refers to a group of normal alkanes with a carbon number generally greater than 15, while oil in this case simply refers to everything else in the system. Both of these two groups of substances contain hundreds, if not thousands, of components. The light components with low carbon numbers such as methane, ethane, propane, etc., are volatile and tend to evaporate when pressure is decreased. On the other hand, the waxy components are much heavier, and they precipitate at temperatures below the WAT. Figure 3.1 shows a representative phase diagram of a typical crude oil.

The thermodynamic modeling of waxes to characterize the wax precipitation from live crude oils composed of N hydrocarbon species involves three-phase (vapor, liquid, and wax) simulation to calculate the following $3N + 3$ quantities:

- The total amount of the vapor phase (n^V)
- The total amount of the liquid phase (n^L)
- The total amount of the solid phases (n^S)
- The molar composition of the vapor phase ($y_1, y_2, ..., y_i, ... y_n$)
- The molar composition of the liquid phase ($x_1, x_2, ..., x_i, ... x_n$)
- The molar composition of the solid phases ($s_1, s_2, ..., s_i, ... s_n$)

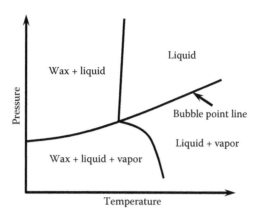

FIGURE 3.1
Typical phase diagram of a waxy crude oil. (From Leontaritis, K. J. *Fuel Sci. Technol. Int'l., 14,* 13–39, 1996.)

Among the $3N + 3$ variables, the amount of the solid phase, n^S, and the solid phase composition, $(s_1, s_2, ..., s_i, ... s_n)$, are of the greatest interest in regard to wax precipitation characterization. Compared to the live crude oils, the stock tank oils (STOs) are usually deficient in light ends due to the gas removal at a separator upstream of the stock tank. Therefore, wax thermodynamic modeling of STO sometimes requires the simulation of the liquid–solid equilibrium, as the gas phase is not present. Figure 3.2 shows the key steps in performing wax thermodynamic modeling.

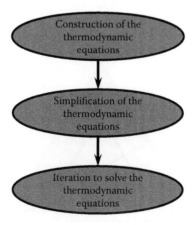

FIGURE 3.2
Flowchart of the steps involved in wax thermodynamic modeling.

The first two steps as shown in Figure 3.2 will be introduced in detail in the remainder of this chapter.

3.3 Step 1 in Thermodynamic Modeling: Construction of the Thermodynamic Equations

$3N + 3$ equations are needed to determine the aforementioned $3N + 3$ variables describing the phase equilibrium. Consequently, the first step in wax thermodynamic modeling is to identify the $3N + 3$ equations to be solved during thermodynamic modeling. All commercial thermodynamic modeling software packages automatically construct these equations. However, these equations are "the content of a black box" to most users. On the other hand, understanding the theory implemented in software packages is essential for one to critically interpret the modeling results. In this section, we present how the equations are constructed based on the first principle of thermodynamics.

The $3N + 3$ equations to be solved during thermodynamic modeling can be divided into three sets: phase equilibrium equations, mass balances, and constitutive equations. The first sets of equations, phase equilibrium equations, are constructed based on the "equal fugacity condition," as illustrated in Figure 3.3.

At thermodynamic equilibrium, the fugacity of species i in the vapor, liquid, and solid phases must equal

$$f_i^{\mathrm{V}} = f_i^{\mathrm{L}} = f_i^{\mathrm{S}} \tag{3.1}$$

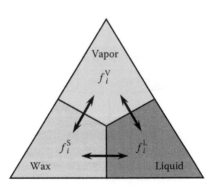

FIGURE 3.3
Three-phase equilibrium among the vapor, liquid, and precipitated wax. The fugacity of each component i in different phases should equal when phase equilibrium is achieved.

The fugacity of species i, f_i^V, f_i^L, and f_i^S, depends on temperature, T, pressure, P, and the molar composition of the phases. Equation 3.2 emphasizes the dependency of f_i^V, f_i^L, and f_i^S on T, P, and the molar compositions:

$$f_i^V = f_i^V(T, P, y_1, y_2, ..., y_i, ... y_n)$$

$$f_i^L = f_i^L(T, P, x_1, x_2, ..., x_i, ... x_n) \qquad (3.2)$$

$$f_i^S = f_i^S(T, P, s_1, s_2, ..., s_i, ... s_n)$$

Equating the fugacities of species in different phases, $f_i^V = f_i^L = f_i^S$, one generates the phase equilibrium equations with respect to the molar compositions in different phases $x_i's$, $y_i's$, and $s_i's$:

$$f_i^V(T, P, y_1, y_2, ..., y_i, ..., y_n) = f_i^L(T, P, x_1, x_2, ..., x_i, ..., x_n)$$
$$\qquad (3.3)$$
$$f_i^L(T, P, x_1, x_2, ..., x_i, ..., x_n) = f_i^S(T, P, s_1, s_2, ..., s_i, ..., s_n)$$

For a mixture of N components, a total of $2N$ equations in the form of Equation 3.3 can be constructed.

The second set of equations, mass balances, are based on the material balance of each component i. The total amount of i in the vapor, liquid, and solid phases must equal the amount of i in the feed:

$$n^V y_i + n^L x_i + n^S s_i = n_i^F \qquad (3.4)$$

Writing the material balance of each of the N components gives a total of N equations in the form of Equation 3.4.

The molar compositions $(y_1, y_2, ..., y_i, ... y_n)$, $(x_1, x_2, ..., x_i, ... x_n)$, and $(s_1, s_2, ..., s_i, ... s_n)$ should also satisfy the following three constitutive equations:

$$\Sigma_i^N y_i = 1$$

$$\Sigma_i^N x_i = 1 \qquad (3.5)$$

$$\Sigma_i^N s_i = 1$$

Combining the $2N$ phase equilibrium equations in the form Equation 3.3, the N material balance equations in the form of Equation 3.4, and the three constitutive equations in the form of Equation 3.5 gives a system of $3N + 3$ equations. These equations can be solved simultaneously to determine the aforementioned $3N + 3$ quantities of interest: n^V, n^L, n^S $y_i's$, $x_i's$, and $s_i's$.

Although the $3N + 3$ equations to be solved have already been identified, the mathematical forms of the phase equilibrium equations in Equation 3.3 have not been discussed yet.

The construction of these phase equilibrium equations will now be discussed.

All existing wax thermodynamic models develop the vapor–liquid phase equilibrium equation based on an equation of state (EOS) of the vapor–liquid mixture. The model development for vapor–liquid fugacity based on the EOS can be readily found in all modern thermodynamic texts (Elliott & Lira, 2012; Sandler, 2006) and thus are not detailed here. However, it is worth pointing out that the EOS mixture parameters used in wax thermodynamic modeling require EOS parameters of each component of the vapor–liquid phases. For example, in the wax thermodynamic model by Lira-Galeana, Firoozabadi, and Prausnitz (1996), the Peng–Robinson EOS is used for fugacity calculation:

$$P = \frac{RT}{V - b_{\text{mixture}}} - \frac{a_{\text{mixture}}}{V(V + b_{\text{mixture}}) + b_{\text{mixture}}(V - b_{\text{mixture}})} \tag{3.6}$$

The mixing rules developed by Chueh and Prausnitz (1967) are used to calculate a_{mixture} and b_{mixture} for the mixture based on the a's and b's of the pure components, and the mixing rules typically involve empirical constants to adjust the model to fit experimental measurements.

Since no EOS is available for the solid phase, the solid–liquid phase equilibrium equation needs to be constructed in a different manner. We will go through a step-by-step procedure for the development of solid–liquid phase equilibrium equations. Firstly, the fugacity of a pure solid is calculated based on the fugacity of a pure liquid and the Gibbs free energy change from pure liquid to pure solid:

$$\ln \frac{f_{i,\text{pure}}^{\text{L}}}{f_{i,\text{pure}}^{\text{S}}} = \frac{\Delta H_i}{RT}\left(1 - \frac{T}{T_i^f}\right) + \frac{\Delta Cp_i}{R}\left(1 - \frac{T_i^f}{T}\right) + \frac{\Delta Cp_i}{R}\ln\frac{T_i^f}{T} + \int_{P_0}^{P} \frac{\Delta V_i}{RT}\,dp \tag{3.7}$$

In a nonideal solid/liquid mixture, the mole fraction of component i needs to be corrected by the activity coefficient to account for the effect of molecular interaction on the chemical potential:

$$f_i^{\text{L}} = x_i \gamma_i^{\text{L}} f_{i,\text{pure}}^{\text{L}}$$
$$f_i^{\text{S}} = s_i \gamma_i^{\text{S}} f_{i,\text{pure}}^{\text{S}} \tag{3.8}$$

Substituting Equation 3.7 into Equation 3.8 yields

$$\ln \frac{f_i^{\text{L}}}{f_i^{\text{S}}} = \ln \frac{x_i \gamma_i^{\text{L}}}{s_i \gamma_i^{\text{S}}} + \frac{\Delta H_i}{RT}\left(1 - \frac{T}{T_i^f}\right) + \frac{\Delta Cp_i}{R}\left(1 - \frac{T_i^f}{T}\right) + \frac{\Delta Cp_i}{R}\ln\frac{T_i^f}{T} + \int_{P_0}^{P} \frac{\Delta V_i}{RT}\,dp$$

$$\tag{3.9}$$

γ_i^L and γ_i^S are functions of the composition of the mixture. Therefore, Equation 3.9 is an equation with the liquid–solid phase compositions x_i and s_i as the independent variables:

$$\ln \frac{s_i \gamma_i^S (s_1, s_2, ..., s_i, ...)}{x_i \gamma_i^L (x_1, x_2, ..., x_i, ...)} = \frac{\Delta H_i}{RT} \left(1 - \frac{T}{T_i^f} \right) + \frac{\Delta Cp_i}{R} \left(1 - \frac{T_i^f}{T} \right)$$

$$+ \frac{\Delta Cp_i}{R} \ln \frac{T_i^f}{T} + \int_{P_0}^{P} \frac{\Delta V_i}{RT} \, dp \qquad (3.10)$$

Different thermodynamic models use different theories to analyze the liquid–solid phase nonidealities and estimate the activity coefficients based on composition. Depending on the complexity of the theory, the mathematical form of Equation 3.10 can be complex, and iterative solution procedures are usually necessary. Some models avoid the mathematical complexities by assuming either the liquid or the solid phase to be an ideal mixture.

Besides the phase nonidealities, the following two issues of n-alkane solid–liquid phase transition also require attention during the modeling of solid–liquid thermodynamic equilibrium:

- n-Alkanes can undergo a secondary phase transition below the melting point.
- Mixtures of n-alkanes can exist in the form of multiple solid phases in equilibrium with each other.

These two phase transition characteristics of n-alkane and their consequence on the modeling wax solid–liquid equilibrium will now be discussed.

The secondary phase transition of n-alkanes is illustrated by an example of the phase transition of pure $n\text{-}C_{23}H_{48}$ as observed in an experimental study using DSC by Dirand et al. (2002). Figure 3.4 shows the enthalpy change during the cooling of $n\text{-}C_{23}H_{48}$.

As indicated by the change of enthalpy, $n\text{-}C_{23}H_{48}$ undergoes a phase transition from liquid to solid at its melting point, ~320 K. The enthalpy of crystallization is released during this phase transition. A secondary phase transition occurs at ~313 K. The enthalpy of secondary phase transition is released during the transition. The cloud point of waxy crude is usually below the secondary phase transition temperature (Coutinho & Stenby, 1996). Therefore, the summation of enthalpy of fusion and the enthalpy of secondary phase transition should be included when calculating the Gibbs free energy change, as shown in Equation 3.11:

$$\Delta H_i = \Delta H_{i,m} + \Delta H_{i,tr} \qquad (3.11)$$

In addition to the phenomena of secondary phase transition, multiple immiscible solid phases with different crystal structures and compositions

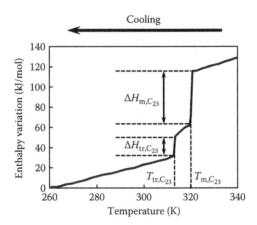

FIGURE 3.4
Change of enthalpy of n-$C_{23}H_{48}$ subject to cooling. (From Dirand, M. et al., *J. Chem. Thermodyn.*, 34, 1255–1277, 2002.)

can coexist below the WAT, further complicating the thermodynamic modeling of the liquid–solid phase equilibrium. Kravchenko's rule of thumb on solid phase miscibility of n-alkanes suggests that total solid phase miscibility between two n-alkanes with different carbon numbers can only occur when the difference in the carbon number is less than 6% the carbon number of the alkanes (Kravchenko, 1946). An experimental study by Snyder, Conti, Strauss, and Dorset (1993), Snyder, Goh, Srivatsavoy, Strauss, and Dorset (1992), and Snyder et al. (1994) confirmed the existence of multiple immiscible solid phases below the WAT. Figure 3.5 shows a phase diagram of a binary mixture of C_{30} and C_{36}.

As can be seen from the phase diagram, a hexagonal solid phase first forms as the temperature decreases. The hexagonal phase starts to transit to orthorhombic phase as the temperature is further decreased. The hexagonal and orthorhombic phases can coexist during this transition depending on the molar composition of the system. At temperatures below approximately 55°C, two orthorhombic phases can coexist with each orthorhombic phase rich in one of the two alkanes.

The existence of multiple solid phases poses further mathematical complexity on wax thermodynamic modeling, as the fugacity of a particular component i in each solid phase is equal, resulting in more phase equilibrium equations to be solved.

As discussed in this section, the following characteristics are important for wax thermodynamic modeling:

- The liquid-phase nonidealities
- The solid-phase nonidealities
- The existence of multiple solid phases

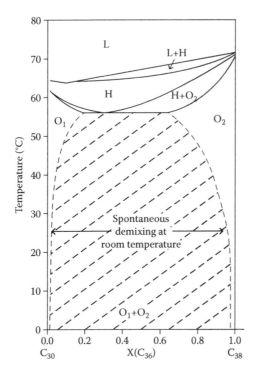

FIGURE 3.5

Phase diagram of a binary mixture of C_{30} and C_{36} with varying concentration of C_{30}: L: liquid; H: hexagonal crystal; O_1: orthorhombic crystal rich in C_{30}; O_2: orthorhombic crystal rich in C_{36}. (From Snyder, R. G. et al., *J. Phys. Chem.*, *96*, 10008–10019, 1992.)

The majority of published wax precipitation models make simplifying assumptions, typically with regard to the ideality of the liquid and solid phases or are limited to a single solid phase. Coutinho (1999) offers a model to predict solid-phase nonidealities using UNIversal QUAsiChemical (UNIQUAC) method and is perhaps the most thermodynamically complete published method.

3.4 Step 2 in Thermodynamic Modeling: Simplification of Thermodynamic Equations

Table 3.1 summarizes the simplifications implemented in various models. This section describes the simplified models including Conoco model, Won's model, Pedersen's model, and Lira-Galeana's model (Erickson, Niesen, & Brown, 1993; Lira-Galeana et al., 1996; Pedersen, Skovborg, & Rønningsen, 1991; Won, 1986). Section 3.4 will be dedicated to describing Coutinho's

TABLE 3.1

Summary of Different Thermodynamic Models' Treatments Regarding Phase Nonidealities and Existence of Multiple Solid Phases

Model	Consider Liquid-Phase Nonideality?	Consider Solid-Phase Nonideality?	Consider Multiple Solid Phases?	Consider Secondary Phase Transition?	Incorporated in Commercial Software Packages?
Conoco model	No	No	Yes, empirically	No	Grand Unified Thermodynamic Simulator (GUTS)
Won's model	Yes, by regular solution theory	Yes, by regular solution theory	No	No	No
Pedersen's model	Yes, by regular solution theory	Yes, by regular solution theory	No	No	PVTsim
Lira-Galeana et al.'s model	Yes, by EOS	No	Yes, by phase stability analysis	No	No
Coutinho's model	Yes, by Flory free volume theory and by UNIQUAC Functional-Group Activity Coefficients (UNIFAC) method	Yes, by Wilson model or by UNIQUAC model	Yes, by UNIQUAC model	Yes	Multiflash

model (Coutinho, 1998, 1999; Coutinho & Daridon, 2001; Coutinho, Edmonds, Moorwood, Szczepanski, & Zhang, 2006; Coutinho & Stenby, 1996; Daridon, Coutinho, & Montel, 2001; Pauly, Daridon, & Coutinho, 2004).

3.4.1 Wax Thermodynamic Models Assuming Single Solid Phase

The earliest attempt of the thermodynamic modeling of wax precipitation dated back to Won's study in the 1980s. Won modeled the liquid–solid equilibrium based on a simplified solid–liquid phase equilibrium equation:

$$\ln \frac{s_i \gamma_i^S}{x_i \gamma_i^L} = \frac{\Delta H_i}{RT}\left(1 - \frac{T}{T_i^f}\right) \tag{3.12}$$

where the activity coefficients γ_i^L, γ_i^S are calculated using the Scatchard–Hildebrand model, as shown in Equation 3.13:

$$\ln \gamma_i = \frac{V_i(\bar{\delta} - \delta_i)^2}{RT} \tag{3.13}$$

Two assumptions, as will be discussed in the following text, were made to simplify the solid–liquid phase in Equation 3.10 to arrive at Equation 3.12. Although the computational intensity is significantly reduced by assumptions, the accuracy of the modeling results is also affected. The validity of these assumptions will now be assessed under different conditions.

Won's model first assumes that the molar volume difference between the liquid and solid phase is small (only 10%), and therefore, the Poynting correction, $\int_{p_0}^{p} \frac{\Delta V_i}{RT}\, dp$, is insignificant in the low-to-moderate pressure range compared to the difference in solid- and liquid-phase standard enthalpy.

The relative magnitude of the contributions to the Gibbs free energy from the enthalpy of phase transition and the Poynting correction will be quantified by considering the following example of the phase transition of n-$C_{20}H_{42}$.

Example

The contribution to Gibbs free energy from the Poynting correction, $\int_{p_0}^{P} \Delta V_i\, dp$, and the contribution from enthalpy change, $\Delta H_i\left(1 - \frac{T}{T_i^f}\right)$, will now be calculated. The material physical properties given in Table 3.2 were collected to carry out these calculations.

TABLE 3.2

Parameters Used for the Calculation of Phase Transition of n-$C_{20}H_{44}$

Material Property	Value
Melting point (T^f)	310 K
Heat of fusion (ΔH)	70 kJ/mol
Change of volume at phase transition (ΔV)	~3.6 × 10^{-2} L/mol
Gas constant (R)	8.314 J/mol/K

At an STO temperature ($T = 298$ K) and pressure ($p = 2$ atm), the contribution from Poynting correction can be calculated as follows:

$$\int_{P_0}^{P} \Delta V\, dp \simeq \Delta V \times (p - p_0) = 3.6 \times 10^{-2} \frac{L}{mol} \times 101\ kPa = 3.6 \frac{J}{mol} \quad (3.14)$$

The contribution from the enthalpy of crystallization, termed $\Delta H_i^f \left(1 - \dfrac{T}{T_i^f}\right)$, can be calculated as follows:

$$\Delta H \left(1 - \frac{T}{T^f}\right) = 70 \frac{kJ}{mol} \times \left(1 - \frac{298\ K}{310\ K}\right) = 2710 \frac{J}{mol} \quad (3.15)$$

The contribution from the Poynting correction, 3.6 J/mol, is only 0.13% of the contribution from the enthalpy of crystallization, 2710 J/mol. However, the contribution from the Poynting correction increases with pressure and can eventually become greater than the contribution from the enthalpy of crystallization. Figure 3.6 shows the comparison of the

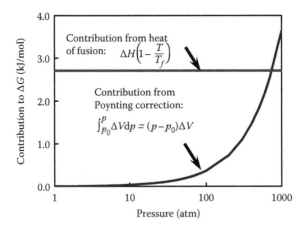

FIGURE 3.6

Comparison between the contribution to Gibbs free energy change in phase transition from heat of fusion and the contribution from the Poynting correction.

contribution from the Poynting correction and the standard enthalpy difference for a phase transition from liquid to solid as a function of pressure.

As can be seen from Figure 3.6, at low-to-moderate pressure (<100 atm), the contribution from Poynting correction can indeed be neglected compared to the heat effect without causing a relative deviation of greater than roughly 15%. The contribution from Poynting correction should not be neglected at pressure greater than approximately 200 atm, which is a typical value for reservoir pressure. Actually, wax thermodynamic modeling at close-to-reservoir pressure is not a trivial task. At elevated pressures, light ends can remain dissolved in the liquid phase and help solvate heavy n-paraffins. Consequently, the WAT of a crude oil containing light end is expected to be lower than the WAT of the same crude without light ends. Wax thermodynamic modeling without considering the effect of light ends often leads to overconservative predictions of the WAT and the WPC.

Besides the assumption of negligible Poynting correction, Won's model also assumes that the difference between the liquid- and solid-phase heat capacity, ΔCp_i, is also insignificant compared to the difference in the standard enthalpy and can be neglected. However, according to the study by Lira-Galeana et al. (1996), models considering the heat capacity difference provide predictions consistent with experimental data, while a 50% relative difference between model predictions and experimental data was observed without considering the change in heat capacity.

Example

Won's model will be used to calculate the amount of precipitated solid at room temperature, i.e., $T = 298$ K, based on a (n-$C_{10}H_{44}$ + n-$C_{24}H_{50}$ + n-$C_{26}H_{54}$) ternary mixture with the composition as shown in Table 3.3.

The goal of the thermodynamic simulation is to determine the number of moles of the solid and liquid phase, n^S and n^L, respectively, as well as the molar composition of each component in the two phases: s_{10}, s_{24}, s_{26}, x_{10}, x_{24}, and x_{26}, where s_i and x_i represent the mole fraction of component i in the solid phase and the liquid phase, respectively. In order to calculate these variables of interest, the physical properties of n-$C_{10}H_{44}$, n-$C_{24}H_{50}$, and n-$C_{26}H_{54}$, including their heats of crystallization and melting point, are needed as input parameters for thermodynamic modeling. These parameters are collected from the National Institute of Standards and Technology (NIST) database and are summarized in Table 3.4.

TABLE 3.3

Composition of the Ternary Mixture Investigated in the Sample Thermodynamic Modeling

Component	Content
n-$C_{10}H_{44}$	0.8 mol
n-$C_{24}H_{50}$	0.1 mol
n-$C_{26}H_{54}$	0.1 mol

TABLE 3.4

Material Properties of n-$C_{10}H_{44}$, n-$C_{24}H_{50}$, and n-$C_{26}H_{54}$ as Input Parameters to the Thermodynamic Model

Material Property	Value
Melting point of n-$C_{10}H_{44}$ (T_{10}^f)	234 K
Melting point of n-$C_{24}H_{50}$ (T_{24}^f)	324 K
Melting point of n-$C_{26}H_{54}$ (T_{26}^f)	330 K
Heat of crystallization of n-$C_{10}H_{44}$ (ΔH_{10})	28.7 kJ/mol
Heat of crystallization of n-$C_{24}H_{50}$ (ΔH_{24})	81.7 kJ/mol
Heat of crystallization of n-$C_{26}H_{54}$ (ΔH_{26})	93.5 kJ/mol

According to the study by Won, the activity coefficients of species in the liquid and solid phases are both equal to 1. Based on this simplification, the phase equilibrium equations can be written as follows:

$$\ln\frac{s_{10}}{x_{10}} = \frac{\Delta H_{10}}{RT}\left(1-\frac{T}{T_{10}^f}\right) \tag{3.16}$$

$$\ln\frac{s_{24}}{x_{24}} = \frac{\Delta H_{24}}{RT}\left(1-\frac{T}{T_{24}^f}\right) \tag{3.17}$$

$$\ln\frac{s_{26}}{x_{26}} = \frac{\Delta H_{26}}{RT}\left(1-\frac{T}{T_{26}^f}\right) \tag{3.18}$$

The material balances on each species can be written as follows:

$$n^L \cdot x_{10} + n^S \cdot s_{10} = n_{10} \tag{3.19}$$

$$n^L \cdot x_{24} + n^S \cdot s_{24} = n_{24} \tag{3.20}$$

$$n^L \cdot x_{26} + n^S \cdot s_{26} = n_{26} \tag{3.21}$$

The mole fractions of each species in the liquid and solid phases should also satisfy the following constitutive equations:

$$x_{10} + x_{24} + x_{26} = 1 \tag{3.22}$$

$$s_{10} + s_{24} + s_{26} = 1 \tag{3.23}$$

Solving Equations 3.16 through 3.23 numerically yields the following results:

$$n^L = 0.822 \text{ mol}$$

$$n^S = 0.178 \text{ mol}$$

$$s_{10} = 0.071$$

$$s_{24} = 0.427$$

$$s_{26} = 0.503 \tag{3.24}$$

$$x_{10} = 0.958$$

$$x_{24} = 0.029$$

$$x_{26} = 0.013$$

As can be seen from these numerical results, the mole fraction of the solid phase at room temperature is 0.178. The liquid phase at room temperature is mainly composed of n-$C_{10}H_{44}$, whereas the solid phase is mainly composed of the two heavy n-paraffins: n-$C_{24}H_{50}$ and n-$C_{26}H_{54}$.

Pedersen et al. improved Won's model by

- Including the terms containing ΔCp_i in the phase equilibrium
- Improving the empirical correlations to estimate physical properties by fitting model predictions with experimental data of 17 North Sea crude oils (Pedersen, Skovborg, & Rønningsen, 1991)

Pedersen's model incorporated the assumption of an ideal solid phase. However, it should be noted that assuming the solid phase to be an ideal mixture can lead to overestimation of the WPC and thus overpessimism about the wax deposition problems in the pipeline. Figure 3.7 shows examples of overestimation of the WPC potentially due to assuming an ideal solid phase.

Therefore, proper modeling of the solid phase nonideality is essential to the accurate prediction of wax precipitation characteristics. Theoretically comprehensive approaches for the modeling of solid-phase nonidealities are introduced in Section 3.4, where Coutinho's model is discussed.

3.4.2 Wax Thermodynamic Models Considering Multiple Solid Phases by Empirical Approaches: The Conoco and Lira-Galeana Models

Different from all the models discussed in Section 3.3.1, the Conoco model considers the existence of multiple solid phases, although in an empirical manner. As temperature drops right below the WAT, the first solid phase of wax forms. As temperature further decreases, new solid phases of wax can form, and the newly formed phase can have a different composition from the early formed phases. The Conoco model assumes that the liquid phase is only at thermodynamic equilibrium with the most recently formed solid phase, while an earlier formed solid phase does not affect the solid–liquid phase equilibrium at the current temperature. Based on this assumption, the Conoco model empirically

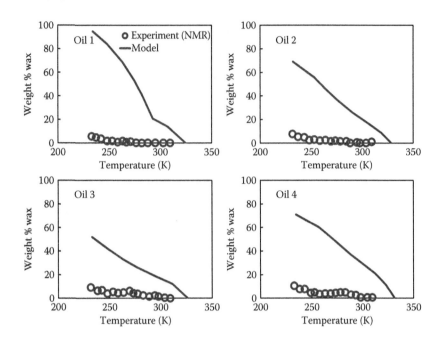

FIGURE 3.7
Overprediction of the WPCs potentially due to the assumption of ideal solid phases. (From Pedersen, K. S. et al., *Energy & Fuels,* 5, 924–932, 1991.)

simulates the existence of multiple solid phases using a staged phase calculation procedure: It calculates the partition of wax between the solid and liquid phase at 1.8°F temperature increments while assuming that the solid phase formed at temperatures greater than $T + 1.8°F$ does not affect the phase equilibrium at T.

Different from the Conoco model, the thermodynamic model by Lira-Galeana et al. treats the formation of multiple pure solid phases using phase stability criteria. Thus, the Lira-Galena et al. and Conoco approaches model precipitation as an expanding core of solids, somewhat analogous to an onion.

3.5 Coutinho's Thermodynamic Model—A Theoretically Comprehensive Thermodynamic Model

Coutinho's model is the only model that considers all of the following perspectives of wax precipitation:

- The liquid-phase nonidealities
- The solid-phase nonidealities
- The existence of multiple solid phases

Coutinho's model calculates the liquid-phase activity coefficients based on the Flory free-volume theory, which accounts for the entropy effects due to the molecule size difference as well as the free-volume effects. The Coutinho model has two variations with different approaches to simulate the solid phase nonidealities: the Wilson model and the UNIQUAC model. The Wilson model calculates the activity coefficients by considering the difference of the interaction between neighboring alkane chains in a pure solid and that in a solid mixture. In a crystal of pure n-C_i-H_{2i+2}, each alkane chain C_i is interacting with alkane chains C_i at adjacent lattice sites through van der Waals forces. In a solid mixture, the adjacent alkane chains to a particular alkane chain might have a different length $j (j \neq i)$. The magnitude of the interaction between a C_i–C_j pair is different from that between a C_i–C_i pair, resulting in a deviation in the Gibbs free energy of i compared to its pure solid phase Gibbs free energy. The calculation of activity coefficients using the UNIQUAC model requires more mathematical computation. However, according to the study by Coutinho et al., the Wilson model and the UNIQUAC model provide a similar prediction of wax precipitation characteristics, as can be seen from Figure 3.8.

A detailed description of the implementation of the UNIQUAC model in modeling the solid-phase nonidealities can be found in the study by Coutinho et al. (2006).

To this point, the theoretical basis of all mainstream wax thermodynamic models has been discussed. These models have all been validated by the thermodynamic modeling of the wax precipitation characteristics of model mixtures and crude oils. Although different models are developed to different extents of theoretical rigor, predictions by each model seem to match experimental measurements in the published work. Therefore, a comparison of the manuscripts does not indicate the level of rigor necessary for accurate predictions. Pauly, Dauphin, and Daridon (1998) conducted a comparative

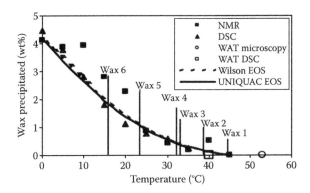

FIGURE 3.8

Comparison between the WPCs by the Wilson model and the UNIQUAC model. (From Coutinho, J. A. P. et al., *Energy & Fuels, 20*, 1081–1088, 2006.)

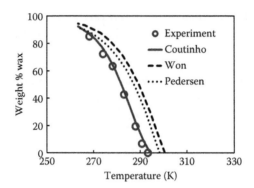

FIGURE 3.9
Comparison of predicted WPC by different wax thermodynamic models with experimental measurements. (From Pauly, J. et al., *Fluid Phase Equilib.*, *149*, 191–207, 1998.)

study covering all mainstream models except the Conoco model and Lira-Galeana et al.'s model based on their predictions of the WPC of the same model mixture. The comparison between different model predictions is shown in Figure 3.9.

As can be seen from Figure 3.9, Coutinho's thermodynamic model provides the most accurate prediction for the WPC for this model mixture. All other models slightly overpredict the WPC potentially due to the assumption of ideal solid or liquid phases.

3.6 Industrial Practice of Wax Thermodynamic Modeling

Thermodynamic models have been developed into commercial software packages, such as GUTS (Conoco model), Multiflash (Coutinho's model), and PVTsim (Pedersen's model). Wax thermodynamic modeling for industrial flow assurance purposes is usually performed using these software packages. In this section, the industrial practice of using the wax thermodynamic model to guide flow assurance consulting is discussed. Figure 3.10 shows a flowchart of performing wax thermodynamic modeling to predict the WAT and the WPCs in software packages.

3.6.1 Fluid Characterization: Preprocessing of Thermodynamic Modeling

In order to use wax thermodynamic models for WAT and WPC prediction, composition of the oil must be input to the thermodynamic model. The oil composition is usually characterized by a testing lab. The following two weight distributions are usually reported:

- *The single carbon number (SCN) distribution*: A SCN fraction with carbon number i includes branched paraffin, cyclic paraffin, aromatics, and n-paraffin with carbon number i. Therefore, the molecular formula of the components in SCN with carbon number i can be written as C_iH_j, where $j = 2i + 2$ represents saturated components and $j \neq 2i + 2$ represents cyclic paraffins, aromatics, and other unsaturated components.

- *The n-paraffin distribution*: the mole/mass fraction of only n-paraffin, $n\text{-}C_iH_{2i+2}$, as a function of carbon number i.

The "SCN distribution" provides information regarding the total amount of hydrocarbon components C_iH_j (including both n-paraffin and non-n-paraffin with the carbon number i) contained by the oil.

Only the n-paraffin ($n\text{-}C_iH_{2i+2}$) portion of component C_iH_j is assumed to precipitate. For a typical crude oil, the non-normal paraffins usually do not precipitate as their melting points are usually significantly lower than the n-paraffin with the same carbon number.

Based on a high-temperature gas chromatography (HTGC) analysis of the crude oil, its SCN distribution and n-paraffin distribution can be obtained by integrating the area under peaks in the gas chromatograph. A different method of integration, e.g., "from baseline" and "from valley to valley" as shown in Figure 3.11, can be used to obtain these two distributions from the gas chromatograph.

It should be noted that the two distributions obtained using different methods of integration and the subsequent thermodynamic calculations using the obtained distributions can be significantly different, as shown in Figure 3.12.

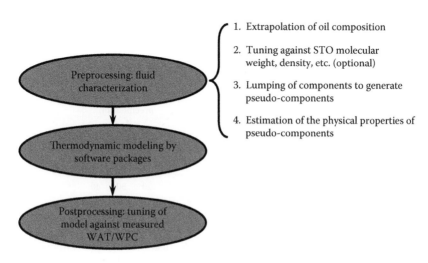

FIGURE 3.10
Flowchart of performing wax thermodynamic modeling using software packages.

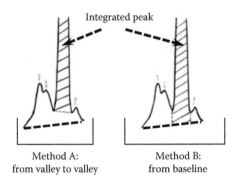

FIGURE 3.11
Two different methods of integration to obtain SCN and n-paraffin distribution from oil–gas chromatograph. (From Coto, B., Coutinho, J. A. P., Martos, C., Robustillo, M. D., Espada, J. J., & Peña, J. L., *Energy & Fuels*, 25, 1153–1160, 2011.)

Different researchers probably follow their own in-house methods when measuring the n-paraffin distribution and the SCN distribution. Figure 3.13 shows these two distributions of a North Sea crude oil provided by Statoil.

Figure 3.13 shows both the mole fraction of the hydrocarbon components and their n-paraffin fractions that were measured up to C_{36}. However, for components with a carbon number greater than 36, the accuracies of the measured mole fractions and n-paraffin fractions become susceptible to experimental errors due to the low absolute amounts of the C_{36+} fractions and the difficulties in determination of high boiling components by GC. Therefore,

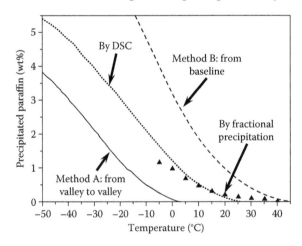

FIGURE 3.12
Comparison of the WPC prediction based on n-paraffin distribution obtained from different methods of integration: from valley to valley and from baseline with experimental measurements by DSC and fractional precipitation. (From Coto, B., Coutinho, J. A. P. et al., *Energy & Fuels*, 25, 1153–1160, 2011.)

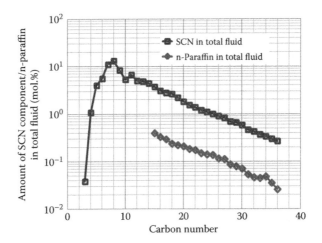

FIGURE 3.13
Composition of the crude oil including n-paraffin distribution of oil S measured by HTGC. (From Zheng, S. et al., *Energy & Fuels*, 27, 7379–7388, 2013.)

the amount of the component with the highest carbon number in an HTGC analysis is usually reported as one number including all C_{36+} components, which is called a plus fraction. The wax precipitation characteristics in terms of the WPC and the composition of the precipitated solid phase are sensitive to the composition of the plus fraction because heavy components are the first to precipitate as temperature is decreased below the WAT (Erickson et al., 1993). Consequently, the mole fractions of the C_{36+}, i.e., components with carbon numbers greater than 36, and the corresponding n-paraffin fractions need to be properly estimated and included in thermodynamic modeling so that model predictions better represent the actual crude oil precipitation characteristics.

Based on field experience, Pedersen, Thomassen, and Fredenslund (1985) suggested extrapolating the SCN and the n-paraffin distribution of the heavy end to a certain limiting carbon number by assuming that the amount of hydrocarbon component C_iH_j and n-paraffin $n\text{-}C_i\text{-}H_{2i+2}$ decreases logarithmically as the carbon number i increases. Pedersen's method of extrapolation has provided satisfactory modeling results for multiple crude oils (Hansen, Fredenslund, Pedersen, & Rønningsen, 1988). A sensitivity analysis showed that for this particular crude oil, including C_{50+} alkane components in thermodynamic modeling affects the predicted precipitation curve by less than 0.1%. Figure 3.14 shows the composition of the oil with the mole fractions of C_{36}–C_{50} components extrapolated using Pedersen's method.

In some other cases, the SCN and the n-paraffin distribution are extrapolated to a carbon number as high as 75 (Erickson et al., 1993). In order for the software packages to automatically extrapolate the plus fraction composition to an appropriate limiting carbon number, additional information such as

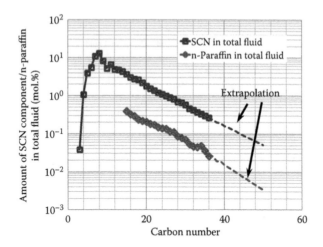

FIGURE 3.14
Extrapolated composition of the crude oil S including SCN distribution and n-paraffin distribution. (From Zheng, S. et al., *Energy & Fuels, 27,* 7379–7388, 2013.)

the plus fraction density and the molecular weight should be supplied to the software packages. If the plus fraction density and the molecular weight are unknown, a "trial-and-error" method suggested by Calsep can be used to adjust the plus fraction molecular weight and density until the flash calculation at stock tank condition predicts the reported STO molecular weight and density.

Modeling the thermodynamic equilibrium of each n-paraffin and non-n-paraffin components requires solving one vapor–liquid and one liquid–solid equilibrium equation for each component. With this crude oil, the compositional analysis resolved the weight distribution over approximately 100 components (both paraffin and nonparaffin). Therefore, approximately 200 equations need to be solved in order to determine the molar composition of each component in each phase. The actual scenario is worse than it already sounds. Each of the 200 equations contains the concentration of components in the same phase. Therefore, each of the 200 equations contains the 100 unknown compositions. Solving this system of equation requires an enormous amount of computer time. In order to reduce computational intensity, components with neighboring carbon numbers are usually lumped to form pseudo-components. Default PVTsim settings lumps real components into 12 pseudo-components (*PVTsim 19 method documentation,* 2009). After lumping, only 24 equilibrium equations instead of some 200 equations need to be solved.

After lumping, each pseudo-component represents a group of real components, and thus, its physical properties should be an average of its constituent real components. As important parameters in the thermodynamic modeling, the averaged critical pressure P_c, the averaged critical temperature T_c, and the averaged acentric factor ω need to be calculated using empirical correlations. Empirical correlations usually estimate P_c, T_c, and ω based on the molecular weight of the pseudo-components.

The preceding three steps

- The extrapolation for the amount of high carbon number components
- The lumping of real components to form pseudo-components
- The calculation of pseudo-component physical properties by empirical correlations

are called the "characterization" of a petroleum fluid. The "characterized" fluid can be passed onto the next subroutine for thermodynamic modeling.

3.6.2 Model Tuning: Postprocessing of Thermodynamic Modeling

Predictions of the WAT and the WPC from software packages should always be carefully evaluated. Multiple uncertainties can exist through the course of performing wax thermodynamic modeling. Some common uncertainties include

- The SCN distribution and the n-paraffin distribution as the input to thermodynamic models are sensitive to the method of integration used to process gas chromatograph.
- The extrapolation of SCN distribution and n-paraffin distribution assumes exponential decay in amounts as the carbon number increases. Some crude oils do not follow this exponential decay.
- The estimation of pseudo-component physical properties might not be accurate.

Model predictions usually need to be benchmarked with experimental measurements. Due to multiple uncertainties, it is rare that the model predicts the exact values of the experimentally measured WAT and WPC. All software packages provide options to adjust certain model parameters (tuning options) in order for the model predictions to match the experimentally measured WAT or wax content. Table 3.5 summarizes the tuning options provided in each software package.

It should be noted that although most model predictions match experimental measurements after tuning, one should remain skeptical about the tuned predictions as all tuning options have obvious shortcomings listed in Table 3.5, because adjustments are made to measured characteristics of oil or wax. Tuned model predictions for WAT and WPC are usually passed on to the next subroutine for wax deposition modeling. Wax deposition modeling will be discussed in Chapter 4.

TABLE 3.5

Summary of Tuning Options in Mainstream Thermodynamic Modeling Software
Packages and Their Shortcomings

Mainstream Software	Tuning Option	Tune to Match	Shortcoming
PVTsim	Wt% of wax in oil	WAT or wax content or both	Fluid might not represent oil composition after tuning
GUTS	Wt% of wax at the precipitation onset	WAT	Inconsistent with lab testing techniques
Multiflash	n-Paraffin heat of fusion	WAT	Inconsistent with the physical properties of the material

3.7 Extended Applications of Wax Thermodynamic Models

The applications of wax thermodynamic models are not limited to predicting the WAT and the WPC. In the determination of the WPC using DSC, one
calculates the WPC based on the DSC trace with an assumed value of 200 J/g
for the wax heat of crystallization. However, using a constant heat of crystallization might not represent the actual wax precipitation characteristics.
Alkane components with the highest carbon number first precipitate when
temperature is decreased right below the WAT. Lighter alkane components
precipitate as temperature is further decreased. The heat of crystallization
of the heavy alkane component and of the light alkane component differ.
Therefore, using a constant value for the heat of crystallization for all the
precipitating components might lead to inaccuracy in the WPC. Wax thermodynamic modeling can be used to improve the accuracy of the calculation
from the DSC trace to obtain the WPC.

In order to achieve a more accurate WPC, the heat of crystallization used
to calculate the amount of solid precipitation should also vary with the temperature in order to account for the fact that alkanes with different heat of
crystallization precipitate at different temperatures. In order to determine
the value of heat of crystallization to be used for calculation, knowledge of
the carbon number of the alkane precipitating at the current temperature is
required a priori. Wax thermodynamic models can predict the carbon number of precipitating alkane at each temperature. Based on this idea, Coto,
Martos, Espada, Robustillo, and Peña (2010) developed an iterative algorithm
using a thermodynamic model. According to Coto et al.'s method, one first
obtains an initial guess for the WPC and oil composition based on the DSC
trace by assuming that

- n-Paraffin n-C_i precipitates at its melting temperature T_i^f.
- The amount of n-C_i precipitated at $T_{m,i}$ equals $w_i = \dfrac{q(T_i^f)}{\Delta H_i^f}$, where $q(T_i^f)$ is the measured heat release at T_i^f.

It should be noted that the temperature at which an alkane component precipitates from a liquid T_i^p only equals its melting temperature T_i^f when it exists as a pure liquid. The precipitation temperature T_i^p of n-C_i from a liquid mixture depends on the composition of the liquid mixture and can be predicted by thermodynamic modeling based on the guessed oil composition. Based on the predicted precipitating temperature (T_i^p), one can recalculate the amount of n-C_i precipitated: $w_i = \dfrac{q(T_i^p)}{\Delta H_i^f}$ at T_i^p, and obtain a second WPC and oil composition. More iterations to obtain the WPC and the oil compositions are performed until no significant change in the WPC and the oil compositions can be observed. Figure 3.15 summarizes this iterative algorithm based on wax thermodynamic modeling to improve the accuracy of WPC obtained from DSC traces.

In addition to the application of thermodynamic modeling of waxes to improve experimental methods of using DSC to measure precipitation curves, the wax thermodynamic model can also be combined with deposition models to predict the wax deposit carbon number distribution. This application will be presented in Chapter 5 together with various applications of wax deposition models.

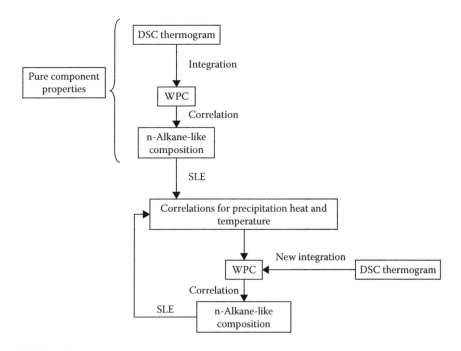

FIGURE 3.15
An iterative algorithm to improve the accuracy of DSC-measured WPC based on thermodynamic modeling. (From Coto, B. et al., *Fuel, 89,* 1087–1094, 2010.)

3.8 Summary

In this chapter, different wax thermodynamic models were introduced. Among these wax thermodynamic models, Coutinho's model is the most rigorous model, while other models implement different assumptions for model simplification. The assumption of ideal solid–liquid phase behavior usually leads to overestimation of the WAT and the WPC.

Wax thermodynamic models can predict the WAT, the WPC, and the composition of precipitated wax. These predictions provide important information regarding wax precipitation characteristics to help assess the possibility and severity of wax deposition issues. It should be noted that in order to obtain reliable predictions from wax thermodynamic models, one should follow certain established procedures. The industrial practice for using wax thermodynamic models was also introduced in this chapter. Reliable WAT and WPC predictions by thermodynamic models can be used as input parameters for wax deposition modeling when the corresponding experimental characterizations are not available. The application of wax thermodynamic models in wax deposition modeling will be introduced in Chapters 4 to 6.

4

Wax Deposition Modeling

In this chapter, the fundamentals of transport phenomena that are used to model wax deposition are discussed in detail. We keep in mind that most mathematical models, no matter how complicated they are, should be based on a description of physical phenomena with certain assumptions. Therefore, before introducing these transport theories, let us first review the two utterly most important questions:

- What is wax deposition?
- How does wax deposition occur?

We will first address these two questions in Section 4.1.

4.1 Wax Deposition Mechanisms

The first question to understand wax deposition would be "What deposits in wax deposition?" Obviously the answer shall be wax, but in which form? We know that there are wax molecules that are dissolved in the oil, and that there are also precipitated wax particles that form a suspension in the oil, as shown in Figure 4.1. Of these two types of wax, which is the source for wax deposition at the pipe wall?

In fact, this question was intensively studied between the 1980s and 2000s. During the early stage of such investigation, the following wax deposit formation mechanisms were initially proposed by Burger et al. (1981):

- *Molecular diffusion*: Wax deposition due to the diffusion of the dissolved *molecules* of the waxy components toward the wall.
- *Shear dispersion*: Wax deposition due to the dispersion of the precipitated *particles* of the waxy components toward the wall.
- *Brownian diffusion*: Wax deposition due to the diffusion of the precipitated *particles* toward the wall. The diffusion of the precipitated particles is caused by Brownian motion.
- *Gravity settling*: Wax deposition due to the settling of the precipitated *particles* of the waxy components toward the bottom of the pipe.

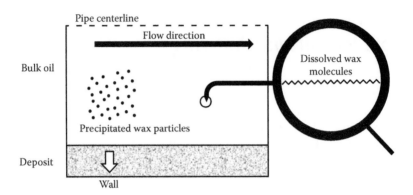

FIGURE 4.1
Schematic of the two types of wax that could potentially cause wax deposition.

We can see that for the first mechanism, molecular diffusion, the subjects of deposition are the dissolved wax molecules in the oil. The depositing materials in the other three mechanisms (shear dispersion, Brownian diffusion, and gravity settling) are the suspended wax particles that have precipitated from the oil. In order to identify which mechanism plays the primary role in wax deposition, researchers have looked at the cause of each of these mechanisms to see if the conditions where these mechanisms would occur are consistent with what happens typically in a pipe flow that is subject to external cooling (which would resemble the transportation of petroleum fluids in the subsea pipelines). And several mechanisms were quickly identified to have insignificant impact on wax deposition: The Brownian diffusion of wax particles is not likely to be valid because the wall temperature is lower than the oil bulk temperature, resulting in more precipitated wax particles at the wall than there are in the bulk oil. Consequently, the impact of Brownian diffusion is to transport these precipitated particles from the wall toward the bulk oil, instead of moving them toward the wall and deposit (Singh et al., 2000). Gravity settling was also believed to be insignificant as there has not been a study claiming that wax deposits are generally thicker at the bottom than at the top of the pipe wall during single-phase oil flow conditions.

The remaining two mechanisms are shear dispersion and molecular diffusion. At early times, the mechanism of shear dispersion was advocated by several researchers (Burger et al., 1981; Hsu & Brubaker, 1995) and was in fact incorporated in early deposition models (such as the wax deposition model in the OLGA industrial flow simulator). However, another early study by Bern et al. (1980) found that the wax deposition rate did not increase with the increasing shear rate of the fluid, casting the first doubt on this mechanism for wax deposition. In fact, many findings in the field of particle fluid dynamics, both theoretical and experimental, seem to suggest that the wax particles in the oil flow are not necessarily to be dispersed toward the wall: The study by Saffman (1965) and Cleaver and Yates (1973) (based on particle fluid mechanics theory) and those by Jimenez et al. (1988), Urushihara et al. (1993), and Garcia et al. (1995) (based on particle velocimetry) all found that particles located in the viscous layer near the wall tend to be re-entrained into the bulk flow by a lifting force generated by the turbulent flow (known as the "Saffman lift force"). These findings suggested that the mechanism of shear dispersion is not significant or even possible. Based on the extensive experimental observations on wax deposition over the past few decades (Hunt Jr, 1962; Eaton & Weeter, 1976; Brown et al., 1993; Creek et al., 1999; Singh et al., 2000; Hoffmann & Amundsen, 2010; Jemmett et al., 2012), it is now generally accepted that the molecular diffusion is the main mechanism for wax deposition. The precipitated wax crystals formed away from the wall do not seem to contribute to the formation of the wax deposit but rather flow with the oil to form a suspension in the fluid mixture. As of today, the mechanism of molecular diffusion has been used by many wax modeling studies (Solaimany Nazar et al., 2001; Roehner & Fletcher, 2002; Hernandez et al., 2003; Venkatesan, 2004; Edmonds et al., 2007; Akbarzadeh & Zougari, 2008; Ismail et al., 2008; Lee, 2008; Merino-Garcia & Correra, 2008; Han et al., 2010; Phillips et al., 2011; Huang et al., 2011a; Lu et al., 2012). Consequently, this mechanism will be discussed in detail in the rest of this chapter.

4.2 Molecular Diffusion as the Main Mechanism for Wax Deposition

A schematic of the mechanism of molecular diffusion is shown in Figure 4.2. The following four steps are involved in this mechanism:

- Step 1: Precipitation of dissolved wax molecules
- Step 2: Generation of radial concentration gradient of dissolved waxy components
- Step 3: Deposition of waxy components on the surface of an existing deposit
- Step 4: Internal diffusion of waxy components in the deposit

These four steps will be discussed in detail subsequently.

4.2.1 Step 1: Precipitation of Dissolved Wax Molecules

Once the fluid temperature decreases to below the wax appearance temperature (WAT), the dissolved waxy components start to precipitate out of the oil

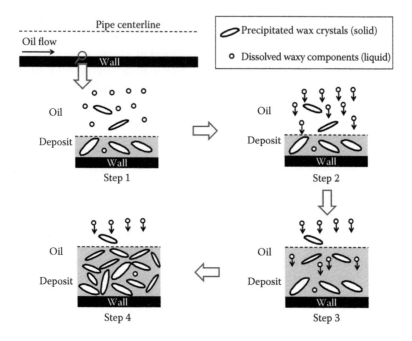

FIGURE 4.2
Schematic of molecular diffusion as the wax deposition mechanism.

and form crystals. Precipitation of the waxy components can occur both at the bulk oil and on the pipe wall as long as the temperature at that particular location is below the WAT, as shown in Figure 4.2 (step 1). As discussed previously, the precipitated wax crystals formed in the bulk are believed to flow with the oil and not deposit on the pipe wall. Therefore, it is the precipitation of the waxy components at the wall that forms the incipient layer of the wax deposit.

4.2.2 Step 2: Formation of Radial Concentration Gradient of Dissolved Waxy Components

During normal cooling conditions, the inner wall usually has a lower temperature than the bulk oil. Therefore, the degree of precipitation of waxy components is generally greater at the wall than in the bulk, resulting in a greater concentration of waxy components in the bulk oil than on the pipe wall, thereby creating a radial concentration gradient of the waxy components between the bulk oil and the wall. The concentration gradient results in the diffusion of the waxy components from the bulk oil, which has a higher concentration of the dissolved waxy components toward the wall, which has a lower concentration of the dissolved waxy components, as highlighted in Figure 4.2. The diffusion coefficient of the waxy components in the oil typically ranges from 10^{-10} to 10^{-9} m^2/s (Hayduk & Minhas, 1982), which is much smaller compared to that for gas systems that are of the order of 10^{-5} m^2/s (Green, 2008).

4.2.3 Step 3: Deposition of Waxy Components on the Surface of an Existing Deposit

As discussed in step 2 (Section 4.3.2), the precipitation of waxy components on the surface of a wall contributes to the formation of wax deposit. Once an incipient deposit layer is formed, the boundary of the oil region becomes the surface of the deposit. In this case, the precipitation of the dissolved waxy components on the deposit surface leads to further growth of the wax deposit, as shown in Figure 4.2 (step 3). Because the waxy crude continues to flow through the pipe, the diffusion of dissolved waxy components toward the deposit continues to occur, resulting in the buildup of the wax deposit.

4.2.4 Step 4: Internal Diffusion and Precipitation of Waxy Components in the Deposit

Although diffusion constantly brings the molecules of waxy components to the oil–deposit interface, not all of these molecules that precipitate at the interface form a new layer of deposit. Some of these dissolved waxy components have been found to continue to diffuse into the wax deposit, resulting in an increase in the wax fraction in the wax deposit. This phenomenon is also known as "deposit aging" and has been observed by numerous researchers (Burger et al., 1981; Lund, 1998; Singh et al., 2000; Singh & Venkatesan, 2001; Hernandez, 2002; Venkatesan, 2004; Hoffmann & Amundsen, 2010; Noville & Naveira, 2012; Bai & Zhang, 2013a,b). This internal diffusion of dissolved waxy components leads to an increase in the wax fraction in the deposit. Consequently, during the course of wax deposition, most of these dissolved waxy components in the deposit are above the solubility limit and could further precipitate to form crystals, resulting in an increase in the solid fraction in the deposit. The phenomena of aging were quantitatively studied by Singh et al. (2000), as shown in Figure 4.3. This study also developed an analysis on the mass balance of the waxy components in the deposit to quantitatively describe the process of aging, which will be discussed in detail in Section 4.4.4.

Let us continue to take a look at the morphology of the wax deposit: The wax deposit consists of a network of precipitated wax crystals, which can entrap the liquid oil and form a porous media. The formation of the network structure was proposed and validated by various researchers (Dirand, Chevallier, & Provost, 1998; Singh et al., 2000; Venkatesan et al., 2005), as shown in Figures 4.4 and 4.5.

As the dissolved waxy components in the deposit continue to precipitate to build up the network structure of the deposit, further internal diffusion of the dissolved waxy components through the entrapped liquid becomes significantly hindered. Consequently, during the process of wax deposition, the phenomenon of aging is usually the most rapid initially and then approaches an asymptotic trend, as observed in Figure 4.3 from the study of Singh et al. (2000).

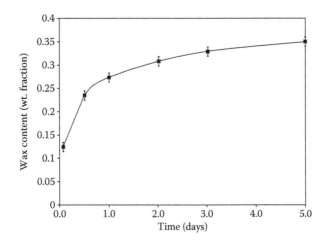

FIGURE 4.3

Evolution of wax fraction in the deposit observed by Singh et al. (2000). The wax fraction in the original oil is 0.0067. (Singh, P., Venkatesan, R., Fogler, H. S., & Nagarajan, N.: Formation and aging of incipient thin film wax-oil gels. *AIChE Journal*. 2000. 46. 1059–1074. Copyright Wiley-VCH Verlag GmbH & Co. KGaA. Reproduced with permission.)

$\bar{n}_c$ = average chain length of the paraffin mixture

FIGURE 4.4

Schematic of the crystal structure in the wax deposit proposed by Dirand et al. (1998) and Singh et al. (2000). (Singh, P., Venkatesan, R., Fogler, H. S., & Nagarajan, N.: Formation and aging of incipient thin film wax-oil gels. *AIChE Journal*. 2000. 46. 1059–1074. Copyright Wiley-VCH Verlag GmbH & Co. KGaA. Reproduced with permission.)

The experimental findings discussed above enable a pathway to develop mathematical models for wax deposition. Based on the mechanism of molecular diffusion, the degree of wax deposition greatly depends on the magnitude of radial diffusion of the dissolved waxy components through the boundary layer of the flow toward the wall. It is generally the most important parameter involved in a wax deposition model and is typically determined by investigating the heat and mass transfer characteristics in the pipe. Such characteristics will be discussed in Sections 4.3 and 4.4.

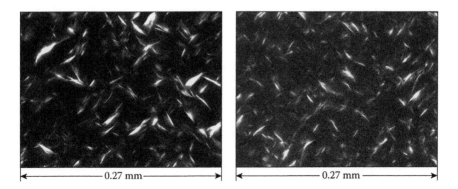

FIGURE 4.5
Microscopic observation of the network structure of a waxy gel by Venkatesan et al. (2005). (Reprinted from *Chemical Engineering Science, 60,* Venkatesan, R., Nagarajan, N. R., Paso, K., Yi, Y.-B., Sastry, A. M., & Fogler, H. S., The strength of paraffin gels formed under static and flow conditions, 3587–3598, Copyright 2005, with permission from Elsevier.)

4.3 Overview of Wax Deposition Modeling

4.3.1 Wax Deposition Modeling Algorithm

A general algorithm for wax deposition modeling is summarized in Figure 4.6.

There are two main calculations involved in this algorithm, and they are highlighted in the dashed boxes in Figure 4.6. The first step involves the hydrodynamic and heat transfer calculations in order to determine the temperature profile along the pipe. The second step involves the mass transfer calculations that are used in combination with the wax precipitation curve to predict the growth rate of wax deposits on the pipe wall. The buildup of the wax deposits on the pipe wall not only reduces the effective diameter of the pipe but also provides additional insulation to reduce heat loss to the ambient environment. Therefore, the heat transfer calculations must be continuously updated until a desired time is simulated. The input parameters for a wax deposition model include two groups of parameters. The first group represents the operating conditions of the pipeline, which often include the dimension, geometry, and insulation of a pipeline; the inlet temperature; the inlet/outlet pressure; the flow rate of the incoming fluids; and the ambient heat transfer conditions. The second group of inputs includes the properties of the fluids, such as density, viscosity, specific heat, thermal conductivity, and phase envelope (for multiphase flow). In terms of wax deposition modeling, the most important property of oil shall be its wax precipitation curve. The wax precipitation curve is generated through

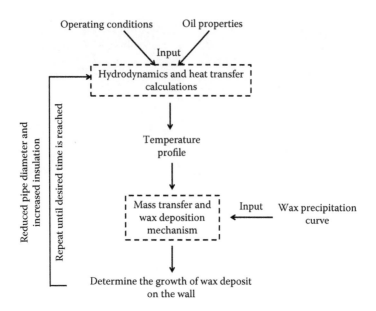

FIGURE 4.6
General algorithm for wax deposition modeling.

thermodynamic studies on the waxy components (n-paraffins) in the oil, as introduced in detail in Chapters 2 and 3. In a deposition model, the wax precipitation curve is used in combination with the temperature/pressure profile in the pipe to determine the radial mass flux of the waxy components toward the wall and eventually calculate the growth rate of wax deposits in the pipe.

4.3.2 Overview of Various Wax Deposition Models

This section provides a detailed comparison on several commercial wax deposition models that have been frequently used by the flow assurance industry, as well as some of the more recently developed academic wax deposition models.

4.3.2.1 Industrialized Commercial Wax Deposition Models

*4.3.2.1.1 Deposition Models of Rygg, Rydahl, and Rønningsen
(1998) and Matzain et al. (2001)*

The modeling works of Rygg et al. and Matzain et al. are currently implemented in an OLGA wax commercial/wax deposition model. OLGA is a multiphase flow simulator that has been developed for several decades and is used widely in the flow assurance industry.

4.3.2.1.2 Deposition Model of Lindeloff and Krejbjerg (2002)

The model study of Lindeloff and Krejbjerg is currently implemented in the DepoWax module in the PVTsim software package. PVTsim is a thermodynamic simulator that has been used extensively for the petroleum industry on fluid characterizations.

4.3.2.1.3 Deposition Model of Edmonds et al. (2007)

FloWax is a wax deposition model developed by Infochem based on the study of Edmonds et al. It should be noted that the equations for heat transfer and wax deposition are originally from Venkatesan (2004).

4.3.2.2 Academic Wax Deposition Models

4.3.2.2.1 Wax Deposition Model from the University of Tulsa

The wax deposition model from the University of Tulsa (TUWAX) is developed through the Tulsa University Paraffin Deposition Projects (TUPDP). One of its early studies was carried out by Matzain (1997). Over the past decade, the TUWAX model has been applied to extensive wax deposition experiments carried out in its in-house flow loop (Lund, 1998; Apte et al., 2001; Hernandez, 2002; Hernandez et al., 2003; Couto, 2004; Bruno, Sarica, Chen, & Volk, 2008). Several versions of TUWAX are available to the members of the TUPDP, and it has been applied in some of the field wax remediation practices (Singh, Lee, Singh, & Sarica, 2011). The information summarized in the study about TUWAX for single phase is based on the study of Hernandez et al. (2003). The information on wax deposition modeling for the gas–oil two-phase flow is based on the study of Apte et al. (2001), while the model's transport equations for the oil–water two-phase flow are likely to be based on the study of Couto (2004) and Bruno et al. (2008).

4.3.2.2.2 Wax Deposition from the University of Michigan (Michigan Wax Predictor)

The Michigan Wax Predictor (MWP) is developed by the University of Michigan Porous Media Research Group. The MWP originated from the study of Singh et al. (2000), which was the first wax deposition model that accounts for the phenomenon of aging (the increase in the wax fraction in the wax deposit). This first version showed remarkable achievement in predicting wax deposition thickness and wax fraction in the deposit for laminar flow conditions. After a decade of doctoral research, the MWP has been extended to turbulent flow conditions and has been well validated by various experimental studies (Venkatesan, 2004; Lee, 2008; Huang, Lee et al., 2011; Senra, Kapoor, & Fogler, 2011; Lu et al., 2012).

In Section 4.4, the theoretical bases and numerical features in the above wax models will be highlighted.

4.4 Detailed Comparison of Different Wax Deposition Models

In this section, we are going to dissect the wax models at a most detailed level. The model configurations as well as the mathematical representations for the fluid flow, heat/mass transfer theories will be thoroughly discussed.

4.4.1 Model Dimensions

The dimensions that a wax deposition model accounts for are often neglected and are actually one of its most important aspects. Because wax deposition occurs as a result of the radial heat loss of the petroleum fluids to the sur-roundings, the radial profiles of temperature and the concentration of the dissolved waxy components are essential for a wax deposition model to accu-rately determine the growth rate of wax deposits. However, solving for the transport characteristics in the radial direction with satisfactory resolution often require significant computational intensity. Such intensity can become a burden for industrial flow simulators, which are mostly discretized only in the axial direction to look for characteristics of the fluids in the bulk flow. An example of the temperature profile during single-phase oil flow is shown in Figure 4.7.

With the current 1-D industrial flow simulators, a single value of tempera-ture is used to represent the bulk oil with another value for the inner wall of the pipe. A laminar sublayer is used to represent the region close to the wall, where the temperature decreases linearly.

For the MWP (an academic wax deposition model), a 2-D (axial and radial) discretization is used, as shown in Figure 4.8. Compared with the approach of 1-D discretization, the 2-D discretization can account for radial variations

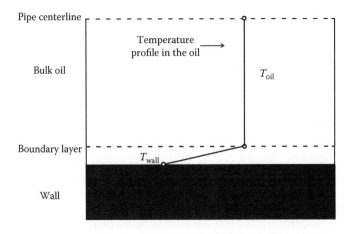

FIGURE 4.7
Discretization of temperature profiles in the 1-D wax deposition models.

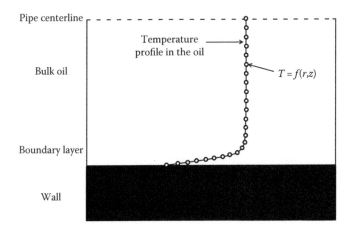

FIGURE 4.8
2-D discretization of temperature profiles in the MWP.

in the oil properties. For example, as temperature decreases in the radial direction of the pipe, the viscosity of the oil can increase, and the diffusion coefficient of the wax in oil can decrease substantially. Using a 2-D discretization in the wax deposition model can help capture these changes to achieve a more accurate production of the heat/mass transfer characteristics.

4.4.2 Hydrodynamics in the Wax Deposition Models

Hydrodynamics can greatly influence the heat and mass transfer characteristics in the pipe especially in multiphase flows. In multiphase flow conditions, one of the most important hydrodynamic parameters is the liquid holdup (the volume fraction of the liquid phase). Because of the drastic difference in the heat capacities between gas and liquid, the liquid holdup has significant impact on the temperature profile in the pipe. A gas–liquid flow with higher liquid holdup generally exhibits a more gradual decrease in temperature during cooling. In order to determine the hydrodynamics of the flow, empirical correlations are usually applied and are summarized in Table 4.1.

For the single-phase flow, empirical hydrodynamic correlations are usually used to determine friction factors in order to calculate the pressure drop across the pipe. These correlations are relatively well established and have been verified by numerous experimental studies (Wilkes, 2005). It should be noted that the correlations for the friction factor are based on a 1-D discretization of the flow simulator. The MWP, due to its 2-D discretization, does not need to determine the friction factor used in hydrodynamic calculations. Instead, it determines the entire radial velocity profile via the correlation of Van Driest (1956) to represent the v^+ and y^+ relationship for turbulent flows and with the parabolic velocity profile for laminar flows.

TABLE 4.1

Summary of Empirical Hydrodynamic Correlations for Different Wax Models

Wax Deposition Models	Matzain (1997); Rygg et al. (1998)	Lindeloff and Krejbjerg (2002)	Edmonds et al. (2007)	Hernandez et al. (2003)	Huang, Lu, Hoffmann, Amundsen, and Fogler (2011)
Software implementation	OLGA wax	PVTsim; DepoWax	FloWax	TUWAX	MWP
Hydrodynamics correlation	Bendiksen, Maines, Moe, and Nuland (1991)	Mukherjee and Brill (1985); Bendiksen et al. (1991)	Brill (1987)	Xiao, Shonham, and Brill (1990); Kaya, Sarica, and Brill (1999)	Van Driest (1956)
Multiphase capability	Yes	Yes	Yes	Yes	Uses averaged properties of fluids as one single-phase fluid

For multiphase flow, empirical correlations are used to estimate velocity, pressure, liquid holdup, and flow regimes. However, due to the complexity in the multiphase flow behavior, uncertainties in these empirical correlations for multiphase flow are much more significant than those for single-phase flow.

4.4.3 Heat Transfer Equations and Correlations

The case of single-phase oil flow will first be considered. Figure 4.9 shows a schematic of the heat transfer characteristics in two dimensions for a cylindrical pipe.

For the 1-D (axial direction) wax deposition models, the general heat transfer equations used for this type of models are shown as follows:

$$\pi R_{pipe}^2 \rho_{oil} C_p U \, dT_{oil} = 2\pi R_{pipe} h_{internal} (T_{oil} - T_{wall}) \, dz \tag{4.1}$$

$$h_{internal} (T_{bulk} - T_{wall}) = h_{external} (T_{wall} - T_{ambient}) \tag{4.2}$$

where
R_{pipe} = the radius of the pipe (m)
ρ_{oil} = density of the oil (kg/m^3)
C_p = heat capacity of the oil (J/K/kg)
U = average velocity of the oil (m/s)

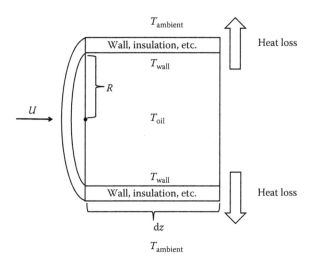

FIGURE 4.9
Schematic of heat transfer characteristics in a single-phase oil flow.

T = temperature (K)
z = axial coordinate (m)
$h_{internal}$ = internal heat transfer coefficient based on the pipe inner diameter (W/m²/K)
$h_{external}$ = external heat transfer coefficient (including the pipe wall, the pipe insulation, and the ambient environment) based on the pipe inner diameter (W/m²/K)

In this case, the temperatures of the bulk oil and at the inner wall, T_{oil} and T_{wall}, are the two variables that are needed to be solved using Equations 4.1 and 4.2. The heat transfer coefficients can be estimated using the empirical correlations summarized in Table 4.2. It should be noted that for multiphase flows, most wax deposition models still use single-phase heat transfer correlations with averaged properties of the phases to estimate the temperature profiles.

The empirical correlation for the internal heat transfer coefficient does not need to be used for the MWP, which has a 2-D discretization scheme. Instead, using the energy balance in both the radial and axial directions (Equation 4.3) with appropriate boundary conditions (Equation 4.4), both the radial and axial temperature profiles can be solved:

$$V_z(r)\frac{\partial T(r,z)}{\partial z} = \frac{1}{r}\frac{\partial}{\partial r}\left[r\left(\varepsilon_{thermal} + \frac{k_{oil}}{\rho_{oil}C_p}\right)\frac{\partial T(r,z)}{\partial r}\right]$$ (4.3)

TABLE 4.2

Summary of Heat Transfer Correlations for Different Wax Deposition Models

	Rygg et al. and Matzain et al. (as Implemented in OLGA Wax)	Lindeloff and Krejbjerg (as Implemented in PVTsim DepoWax)	Edmonds et al. (as Implemented in FloWax)	TUWAX	MWP
Heat transfer correlation	Sieder and Tate (1936), Gnielinski (1976), or Dittus and Bölter (1985)			Sieder and Tate (1936) (single-phase or oil–water flow with averaged properties of fluids) Kim et al. (1999) (Gas–liquid, flow regime–dependent)	Deen (1998) on eddy thermal/ mass diffusivities (single-phase)
Multiphase capability	Uses averaged properties of fluids as one single-phase fluid				Uses averaged properties of fluids as one single-phase fluid

$$\begin{cases} \dfrac{\partial T}{\partial r} = 0, \quad \text{at } r = 0 \\[2ex] h_{\text{external}}(T_{\text{ambient}} - T_{\text{wall}}) = k_{\text{oil}}\dfrac{\partial T}{\partial r}, \quad \text{at } r = R_{\text{pipe}} \end{cases} \tag{4.4}$$

Here, $r(m)$ is the radial coordinate; $\varepsilon_{\text{thermal}}$ (m^2/s) is the eddy thermal diffusivity, which is determined by the von Kármán correlation (Wilkes, 2005); and k_{oil} (W/m/K) is the thermal conductivity of the oil. The details of all the transport equations in this model have been reported in the study of Huang, Lee et al. (2011).

4.4.4 Mass Transfer and Deposit Growth Rate Calculations for the Wax Deposition Models

The previously discussed hydrodynamic and heat transfer calculations provide us the temperature profile of the pipe. In this section, this temperature profile is further utilized to determine the mass transfer characteristics with the wax precipitation curve, which eventually help determine the growth rate of wax deposits.

4.4.4.1 Mathematical Representation of the Molecular Diffusion Mechanism

As previously discussed in Section 4.2, molecular diffusion is the major mechanism for wax deposit formation (Singh et al., 2000). In this mechanism, the precipitated wax particles in the bulk oil do not contribute to wax deposition on the pipe wall. Instead, it is the radial diffusion of dissolved wax molecules that results in the buildup of a wax deposit. The magnitude of diffusion can be described in terms of radial mass fluxes (usually with the unit of kg/m²/s), which are shown as J_A and J_B in Figure 4.10. These two parameters can be further related to the deposition process based on a series of mass balance equations, as shown in Equations 4.5 and 4.6, that are based on the study of Singh et al.

Mass balance at the oil–deposit interface is given by

$$\rho_{\text{deposit}} F_{\text{wax}} \frac{d\delta_{\text{deposit}}}{dt} = (J_A - J_B) \tag{4.5}$$

and mass balance over the entire deposit is given by

$$\frac{\rho_{\text{deposit}} \left(R_{\text{pipe}}^2 - (R_{\text{pipe}} - \delta_{\text{deposit}})^2 \right)}{2\pi(R_{\text{pipe}} - \delta_{\text{deposit}})} \frac{dF_{\text{wax}}}{dt} = J_B \tag{4.6}$$

where ρ_{deposit} (kg/m³) represents the density of the deposit; F_{wax} is the wax mass fraction (the summation of all the waxy components) in the wax deposit (both dissolved in the entrapped oil and precipitated as crystals); δ_{deposit} (m) represents deposit thickness; R_{pipe} (m) represents the radius of the pipe; and t represents time (s).

Equation 4.5 represents a mass balance on dissolved waxy components at the oil–deposit interface, whereas Equation 4.6 is the mass balance on the dissolved waxy components in the deposit, with a radially averaged value of wax fraction in the deposit, F_{wax}. From these balance equations, it can be seen

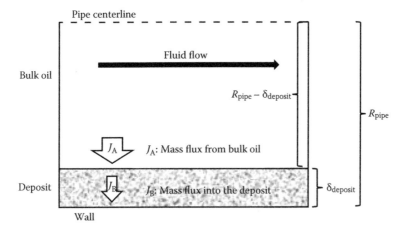

FIGURE 4.10
Schematic of the mass fluxes of the dissolved waxy components.

that the difference of the two mass fluxes ($J_A - J_B$) corresponds to the growth of the deposit layer, while the internal diffusion mass flux, J_B, results in an increase in the wax fraction within the deposit. The increase in the wax fraction in the deposit with time, dF_{max}/dt, is also known as the aging rate of the wax deposit, which is physically related to the hardening of the wax deposit as time increases.

4.4.4.2 Simplification of Wax Deposition Mechanisms

In most of the commercial wax deposition models, the phenomenon of aging (the increase in the wax fraction with time) *is not* accounted for, and the mass flux into the deposit (J_B in Equation 4.6) is neglected. In this case, the two mass balance Equations 4.5 and 4.6 reduce to one equation, as shown in Equation 4.7, where a constant value is used for the wax fraction in the deposit, F_{wax}:

$$\rho_{deposit} F_{wax} \frac{d\delta_{deposit}}{dt} = J_A \tag{4.7}$$

In industrial wax deposition models, the term "porosity of wax deposit" is commonly used instead of the wax mass fraction, F_{wax}. The porosity of the wax deposit refers to the volume fraction of the entrapped oil in the wax deposit. Two approximations are made to relate these two parameters. First, the densities of the oil and the wax are similar, and therefore, the volume fraction of the entrapped oil is assumed to be the same as the mass fraction of the entrapped oil. Second, as the wax deposit is expected to be highly supersaturated (with a wax mass fraction, F_{wax} as high as 50%–90%), significant precipitation is expected to occur in the deposit for the waxy components. Consequently, the amount of dissolved waxy components in the entrapped oil of the deposit is neglected. In other words, it is often assumed that waxy components exist only in the solid phase, and the entrapped liquid phase consists of only nonwaxy components. As a result, the relationship between the porosity and the wax fraction of the deposit is simplified, as shown in the following equation:

$$F_{wax} = 1 - \phi_{deposit} \tag{4.8}$$

where $\phi_{deposit}$ is the porosity of the wax deposit. Consequently, the mathematical representation of molecular diffusion neglecting aging is shown in the following:

$$\rho_{deposit} (1 - \phi_{deposit}) \frac{d\delta_{deposit}}{dt} = J_A \tag{4.9}$$

The method to determine J_A in order to predict the thickness of the wax deposit will now be discussed.

4.4.5 Determining the Mass Flux

The next key step in a deposition model is to determine the value of the mass flux to the interface, J_A, which is usually determined based on Fick's law of diffusion. For the 1-D wax deposition models (as introduced previously in Section 4.4.1), the decrease in the concentration of the waxy components is assumed to be linear within a thin boundary layer near the wall. This linear concentration profile indicates that mass transfer in this thin film occurs through diffusion (rather than convection). A schematic of this profile is shown in Figure 4.11. In this case, J_A can be described as in the following equation:

$$J_A = -D_{wax}\frac{dC}{dr} = D_{wax}\frac{C_{oil} - C_{wall}}{\delta_{mass\ transfer}} = D_{wax}\frac{C_{oil}(T_{oil}) - C_{wall}(T_{wall})}{\delta_{mass\ transfer}} \quad (4.10)$$

where
   C = the concentration of dissolved waxy components (kg/m³)
   D_{wax} = the diffusion coefficient of wax in oil (m²/s)
   T = temperature (K)
   $\delta_{mass\ transfer}$ = the thickness of the mass transfer layer (m)

The configurations of several key parameters in Equation 4.10 are further discussed in Section 4.4.5.1.

4.4.5.1 Diffusion Coefficient of Wax in Oil, D_wax

For all the wax models in this study, the diffusion coefficient of wax in oil, D_{wax}, is determined by the correlation of either Hayduk and Minhas (1982) or Wilke and Chang (1955), as shown in the following equations:

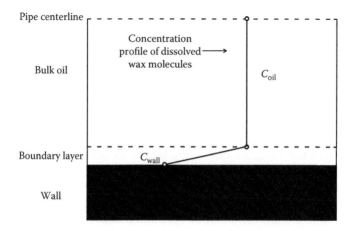

FIGURE 4.11
Schematic of concentration profile for the 1-D wax deposition models.

$$\text{Hayduk–Minhas} \quad D_{wax} = A_{HM} \frac{T^{1.47}\mu_B^{\gamma}}{V_A^{0.71}}, \quad \gamma = \frac{10.2}{V_A} - 0.791 \qquad (4.11)$$

$$\text{Wilke–Chang} \quad D_{wax} = B_{WC} \frac{(\phi_B M_B)^{0.5}T}{\mu_B V_A^{0.6}} \qquad (4.12)$$

where the symbols have the following physical representations:
V_A = the averaged molar volume of n-paraffins (cm³/mol)
μ_B = the solvent viscosity (cP)
ϕ_B = the association parameter for the solvent B (typically used as 1)
M_B = the solvent molecular weight (g/mol)
T = the fluid temperature (K)

The values of the diffusion parameters A_{HM} and B_{WC} depend on the unit of the diffusion coefficient. When the diffusion coefficient has a unit of m²/s, the default values for A_{HM} and B_{WC} in these two correlations are 13.3 × 10⁻¹² and 7.4 × 10⁻¹², respectively. If the diffusion coefficient has a unit of cm²/s, the default values of A_{HM} and B_{WC} becomes 13.3 × 10⁻⁸ and 7.4 × 10⁻⁸ respectively. However, it should be noted that this correlation is originally designed based on binary dilute solutions. In some wax deposition studies of petroleum fluids with a polydispersed composition, A and B become adjustable parameters that are to be benchmarked by lab-scale deposition experiments (Kleinhans, Niesen, & Brown, 2000).

4.4.5.2 Difference in the Concentration of Dissolved Waxy Components in the Bulk Oil and at the Wall, (C*oil* – C*wall*)

The concentration difference ($C_{oil} - C_{wall}$) represents the mass transfer driving force and is essential to wax deposition modeling. In most of the industrial wax models, thermodynamic equilibrium is often assumed to calculate the concentrations of the dissolved waxy components. In other words, the wax solubility curve is used to represent the wax concentrations in the bulk and at the wall. The wax solubility can be determined by inverting the wax precipitation curve, as highlighted in Figure 4.12. This transformation is based on the fact that the amount of wax that remains soluble in the liquid should be equal to the total wax content minus the amount of wax that has precipitated. In this case, a value of the total wax content is needed for this transformation and is usually the amount of wax precipitation at 0°C or 5°C. It should be noted that this value only serves as a reference and should not affect the result of ($C_{oil} - C_{wall}$) because it is canceled during the subtraction.

From the above analysis, it can be concluded that the precipitation curve of wax provides the essential link between heat transfer and mass transfer for wax deposition. It attributes the reason of the concentration difference

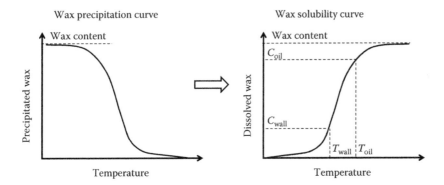

FIGURE 4.12
Conversion from the wax precipitation curve to the wax solubility curve.

($C_{oil} - C_{wall}$) to the temperature difference ($T_{oil} - T_{wall}$) between the bulk oil and at the wall. The temperature difference is known to be the thermal driving force for wax deposition. In many wax deposition studies, this thermal driving force is used as the indicator to cross-compare deposition experiments at different temperatures (Creek et al., 1999; Singh et al., 2000; Paso & Fogler, 2004; Jennings & Weispfennig, 2005). However, it should be noted that two assumptions are made in these analyses using the thermal driving force:

1. The solubility curve of wax is linear in the range of interest.
2. The wax solubility concentrations are given by their thermodynamic equilibrium values.

It can be seen that these two conditions are not generally valid, and thus the use of thermal driving force often has its limitations. Such limitations will be discussed in detail in Chapter 6. As to the second assumption, some of the academic studies have tried to incorporate precipitation kinetics to their models. For example, the TUWAX model includes an additional kinetic precipitation step on the wall, so that the concentration at the wall is not necessary at thermodynamic equilibrium, depending on the value of the surface precipitation rate constant in the study of Hernandez et al. (2003). However, in the MWP, instead of applying precipitation kinetics on the pipe wall, it is applied on the boundary layer of the fluid, where the oil is expected to experience drastic cooling while flowing in the pipeline. This implementation was achieved by using a 2-D mass transfer equation (axial advection and radial diffusion), as shown in the following equation:

$$V_z \frac{\partial C}{\partial z} = \frac{1}{r} \frac{\partial}{\partial r} \left[r(\varepsilon_{mass} + D_{wax}) \frac{\partial C}{\partial r} \right] - k_{precipitation} \left(C - C(eq) \right) \qquad (4.13)$$

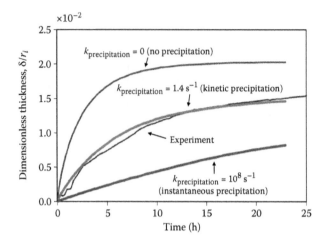

FIGURE 4.13
The impact of wax precipitation kinetics in the oil flow on wax deposition predicted by the MWP in the study of Huang, Lee et al. (2011).

It is seen that the second term on the right-hand side of Equation 4.13, a kinetic precipitation term $k_{precipitation}(C - C(eq))$, is used to account for the kinetics of the precipitation of the dissolved wax molecules when their concentration is above the solubility limit (Lee, 2008). Depending on the value of the kinetic precipitation rate constant, $k_{precipitation}$, the concentration of dissolved wax molecules in the boundary layer can follow thermodynamic equilibrium ($k_{precipitation} \rightarrow \infty$) or slow precipitation ($k_{precipitation} \rightarrow 0 \text{ s}^{-1}$). As the precipitated wax crystals are found to most likely flow with the oil instead of depositing on the pipe wall, a higher precipitation rate would likely reduce the amount of dissolved waxy component in the boundary layer potentially available for diffusion toward the pipe wall. Consequently, a greater precipitation rate in the boundary layer actually leads to less deposition on the pipe wall, as the precipitated wax crystals often flow with the oil instead of depositing on the pipe wall. A comparison with various experimental studies showed that a reasonable value of $k_{precipitation}$ can be on the order of 1.4 s^{-1} (Huang, Lee et al., 2011) and is shown in Figure 4.13.

4.4.5.3 Thickness of the Mass Transfer Layer, $\delta_{mass\ transfer}$

This parameter has arguably caused the most confusion in wax deposition modeling, and different wax deposition models adopt various configurations, some of which are not necessarily correct. In this section, this concept will be explained in detail and differentiated from other concepts. First, the following parameters are introduced:

- Diffusion-equivalent mass transfer layer, $\delta_{\text{diffusion}}$
- Conduction-equivalent mass transfer layer, $\delta_{\text{conduction}}$

4.3.5.3.1 Diffusion-Equivalent Boundary Layer

From a mass transfer perspective, the radial mass flux shown in Equation 4.10 can be expressed with the mass transfer coefficient ($k_{\text{mass transfer}}$), a mass transfer layer thickness ($\delta_{\text{mass transfer}}$), and the Sherwood number (Sh), as shown by the following equation:

$$J_A = D_{\text{wax}} \frac{C_{\text{bulk}} - C_{\text{wall}}}{\delta_{\text{mass transfer}}} = k_{\text{mass transfer}}(C_{\text{bulk}} - C_{\text{wall}}) = \frac{\text{Sh}}{d} D_{\text{wax}}(C_{\text{bulk}} - C_{\text{wall}}) \quad (4.14)$$

where d is the diameter of the pipe (Venkatesan & Folger, 2004).

One can see that a linear concentration profile is used in Equation 4.14 to represent the concentration gradient at the wall. Such a linear concentration profile indicates that only diffusion (rather than advection) is assumed for mass transfer. However, in reality, both advection and diffusion exist in normal pipe flow conditions. Therefore, this approach simply uses a "diffusion-equivalent linear layer" to represent the mass transfer characteristics in the radial direction within the pipe. In this case, the thickness of this type of mass transfer layer is referred to as the diffusion-equivalent mass transfer layer, $\delta_{\text{diffusion}}$. From Equation 4.14, one can see that it is simply the diameter of the pipe divided by the Sherwood number:

$$\delta_{\text{mass transfer}} = \delta_{\text{diffusion}} = \frac{d_{\text{pipe}}}{\text{Sh}} \quad (4.15)$$

It should be noted that this concept of diffusion-equivalent layer should not be mistaken as the "mass transfer boundary layer," $\delta_{\text{mass boundary layer}}$. The mass transfer boundary layer usually refers to the layer within which the wax concentration experiences a 99% decrease (also known as the "99% boundary layer").

4.4.5.3.2 Conduction-Equivalent Boundary Layer

It should be noted that Equations 4.14 and 4.15 do not account for wax precipitation because wax solubility is not involved in these two equations. This assumption is generally not correct because wax does precipitate when the temperature is above its saturation point. Another approach assuming complete thermodynamic equilibrium converts the concentration difference to two elements: (1) the combination of temperature difference, $T_{\text{bulk}} - T_{\text{wall}}$, and (2) the gradient of the solubility curve at the wall temperature, $\left.\dfrac{dC}{dT}\right|_{T_{\text{wall}}}$, by

Equation 4.16. In this case, it is assumed that all the supersaturated wax molecules precipitated instantaneously in the oil:

$$J_A = D_{wax} \frac{dC}{dT}\bigg|_{T_{wall}} \frac{T_{bulk} - T_{wall}}{\delta_{thermal}} \tag{4.16}$$

In Equation 4.17, $\delta_{thermal}$ represents the thickness of the layer where the decrease in temperature can be assumed to be linear. Using a similar derivation to the "diffusion-equivalent layer" shown previously, $\delta_{thermal}$ can be referred to as the "conduction-equivalent layer." Analogous to Equation 4.15, $\delta_{thermal}$ is equal to the pipe diameter divided by the Nusselt number, as shown in the following equation:

$$\delta_{thermal} = \frac{d_{pipe}}{Nu} \tag{4.17}$$

Similar to the clarification made previously, one would like to point out the fact that this conduction-equivalent layer should not be mistaken as the "thermal boundary layer," as the thermal boundary layer often refers to the layer where the temperature experiences a 99% decrease.

Comparing Equations 4.15 and 4.17, it can be seen that the conversion from the diffusion-equivalent layer to the conduction-equivalent layer is merely the ratio between the Sherwood number and the Nusselt number, as shown in the following equation:

$$\frac{\delta_{thermal}}{\delta_{mass}} = \frac{Sh}{Nu} \tag{4.18}$$

Using the correlation of Dittus and Boelter (1985) and the Chilton–Colburn analogy between heat and mass transfer in turbulent flows (Chilton & Colburn, 1934), the Sherwood and Nusselt numbers can be replaced by the Reynolds number, the Schmidt number (for mass transfer), or the Prandtl number (for heat transfer), as shown in the following equations:

$$Nu = 0.023 Re^{0.8} Pr^{0.3} \tag{4.19}$$

$$Sh = 0.023 Re^{0.8} Sc^{0.3} \tag{4.20}$$

Combining Equations 4.18 through 4.20, it is seen that the conversion from the conduction-equivalent layer thickness to the diffusion-equivalent layer thickness can be simply made via the Lewis number, which depends only on the properties of the oil:

$$\frac{\delta_{\text{thermal}}}{\delta_{\text{mass}}} = \frac{\text{Sh}}{\text{Nu}} = \left(\frac{\text{Sc}}{\text{Pr}}\right)^{0.3} = \text{Le}^{0.3} = \frac{k_{\text{oil}}}{\rho_{\text{oil}}C_p D_{\text{wax}}} \qquad (4.21)$$

Again, for the diffusion-equivalent layer, there is no precipitation in the boundary layer (complete supersaturation), and the conduction-equivalent layer represents absolute thermodynamic equilibrium (instantaneous precipitation). Since the reality would most likely fall between these two extremes, most wax deposition models have both configurations (the use of the conduction-equivalent layer and the diffusion-equivalent layer) available for users to establish the magnitude of uncertainties using these two assumptions (Hernandez et al., 2003; Edmonds et al., 2007). For the MWP, the use of a precipitation rate constant, $k_{\text{precipitation}}$, in Equation 4.13 allows one to quantitatively express the effect of kinetic precipitation (a middle ground between these two extremes) on wax deposition. When $k_{\text{precipitation}} \rightarrow \infty$ (rapid precipitation in the boundary layer), it resembles the approach of using the conduction-equivalent layer; when $k_{\text{precipitation}} \rightarrow 0$ (slow precipitation), it resembles the approach of the diffusion-equivalent boundary layer. Because the precipitated wax particles in the oil flow do not tend to adhere to the wall to form wax deposits, and it is the diffusion of the dissolved molecules of the waxy components toward the wall that contributes to the growth of wax deposits, a higher rate of wax precipitation in the bulk would actually reduce the amount of dissolved waxy components available for deposition. Consequently, as $k_{\text{precipitation}}$ increases, the deposition rate actually decreases. The actual value of $k_{\text{precipitation}}$ is often benchmarked through laboratory deposition experiments before application to the field (Huang, Lee et al., 2011). A model prediction with different values of $k_{\text{precipitation}}$ is shown previously in Figure 4.13.

Finally, it should be noted that the diffusion-equivalent layer and the conduction-equivalent layer should not be confused with the laminar sublayer of the flow, as it seems to be the case in certain industrial wax deposition models (Rygg et al., 1998; Lindeloff & Krejbjerg, 2002). The concept of laminar sublayer is used in hydrodynamics (rather than heat and mass transfer) for turbulent flow representing the region near the pipe wall, where the flow exhibits laminar behavior (in terms of fluid mechanics, the region with a dimensionless radial distance from the wall, $y^+ < 5$). This concept is physically irrelevant to the conduction-equivalent layer and the diffusion-equivalent layer, and it should not be used to determine the mass flux for wax deposition modeling. Wax models that use the laminar sublayer thickness usually have to apply an adjustment multiplier in an attempt to address this issue.

4.5 Summary

In this chapter, various aspects of wax deposition modeling are introduced, and we start with the mechanism of wax deposition. Despite multiple mechanisms proposed in the 1980s, it is now generally believed that molecular diffusion is the primary mechanism for wax deposition. Based on this mechanism, wax particles that precipitated in oil do not adhere to the wall and form a deposit. Instead, wax deposition is mainly caused by the radial diffusion of dissolved molecules of waxy components in the oil boundary layer toward the wall (or toward the deposit surface once a layer of deposit has formed). After the arrival of these dissolved waxy components at the deposit surface, they can either precipitate on the surface (resulting in the further increase in the deposit thickness) or continue to diffuse into the deposit and eventually precipitate to form a network structure (leading to the hardening of the deposit).

The mechanism of molecular diffusion has been widely used in many industrial and academic wax deposition models. These models utilize the input of operating conditions as well as the oil properties. Among these oil properties, the most important parameter is the wax precipitation curve, which is often obtained from the output of a wax thermodynamic model. With these two pieces of information, a wax deposition model predicts the rate of the deposition on a pipe wall by performing the following transport calculations:

- Hydraulic and heat transfer calculations
- Mass transfer calculations and the application of the molecular diffusion mechanism

Most of the wax deposition models apply empirical correlations for transport calculations. While different wax deposition models apply various transport correlations, the main goal among these models is too similar, which is to determine the radial mass flux of wax molecules, J_A.

$$J_A = -D_{wax}\frac{dC}{dr} = \begin{cases} D_{wax}\dfrac{C_{oil}-C_{wall}}{\delta_{mass\ transfer}} = D_{wax}\dfrac{C_{oil}(T_{oil})-C_{wall}(T_{wall})}{\delta_{diffusion}} \\[2mm] \text{or} \\[2mm] D_{wax}\dfrac{dC}{dT}\left(-\dfrac{dT}{dr}\right) = D_{wax}\dfrac{dC}{dT}\dfrac{T_{oil}-T_{wall}}{\delta_{conduction}} \end{cases} \qquad (4.22)$$

The key parameters in the mass flux, J_A include the diffusivity of wax in oil, the concentration difference, as well as the mass transfer thickness. Due to the uncertainties associated with these parameters, wax deposition models are often benchmarked to identify the optimum model configurations for field predictions.

5

Introduction to Wax Deposition Experiments

5.1 Importance of Experimental Applications

The ultimate goal of wax deposition modeling is to provide predictions on the possibility and severity of wax deposition in field pipelines in order to establish effective prevention and remediation strategies. However, due to the many simplifications made in wax models, it has become a standard industrial practice to benchmark the wax deposition models with lab-scale or pilot-scale wax deposition experiments prior to their application in the field. Compared to wax deposition in the field, lab-scale wax deposition experiments are often carried out in a better controlled environment with more convenient measurement and characterization techniques. Well-designed and -executed wax deposition experiments not only serve as base cases for model benchmarking but also can provide important information for wax deposition theoretical research and development.

5.2 Wax Deposition Flow Loop

Although arguably the most expensive wax deposition apparatus, a flow loop is often considered the best experimental tool for benchmarking wax deposition models (Creek, Lund, Brill, & Volk, 1999; Hernandez, 2002; Hoffmann & Amundsen, 2010). The flow field in a flow loop is similar to that in field pipelines, and thus the scale-up from flow loop tends to be more reliable than that from other deposition apparatus. A flow loop includes several main components: conditioning system, pumping system, test section, and deposit characterization devices. An example is shown in Figure 5.1.

5.2.1 Conditioning System and Pump System

A storage tank is used to contain the testing waxy oil, and the pumps are used to circulate the oil as well as the coolant during the experiments. During the

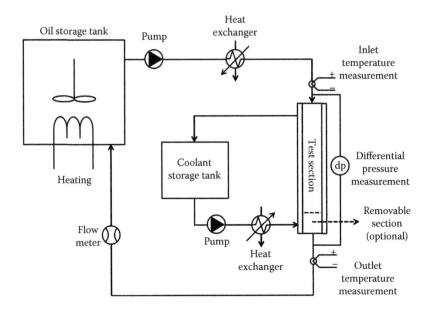

FIGURE 5.1
Schematic of a typical flow-loop wax deposition apparatus.

course of wax deposition, the concentration of the waxy components in the oil decreases as wax deposits in the test section. Therefore, in order to maintain a consistent wax content in the fluid, the volume of the oil storage tank should be sufficient so that the impact of depletion is negligible, as shown in Figure 5.2. An example of quantifying the effect of wax deposition and estimating the appropriate size for the storage tank can be found from the

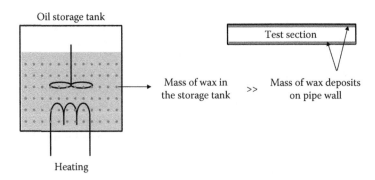

FIGURE 5.2
Principles of estimating the depletion effect to determine the appropriate size of the oil storage tank for wax deposition experiments.

studies of Singh, Venkatesan, Fogler, and Nagarajan (2000) and Hoffmann and Amundsen (2010).

Prior to wax deposition tests, the storage tank oil needs to be homogenized to ensure that the composition of the oil being circulated is representative of the oil that is taken from the field. This goal can be achieved by heating the storage tank oil to well above its wax appearance temperature and at the same time agitating the storage tank oil with a stirring device.

The selection of the oil pumps should be based on the desired Reynolds number (Re) for the oil flow in the deposition experiments. A flow meter is installed in the oil pipe to measure the oil flow rate of the fluid. In order to select the appropriate pump for the coolant, heat transfer calculations should be carried out a priori in order to determine the coolant flow rate that is necessary to achieve the desired wall temperatures.

5.2.2 Test Section

The test section is the most important part of a flow loop where wax deposition is designed to take place. In the test section, oil is cooled by a cooling annulus. Water is often selected as the coolant due to its high heat capacity (Hoffmann & Amundsen, 2010). In some other studies, water is mixed with glycol to decrease its freezing point in order to increase its cooling capacity (Hernandez, 2002). The inlet temperature of the coolant is often controlled by a chiller.

A flow development section should be placed upstream of the test section to allow the velocity profile to be fully developed. During the course of wax deposition, the differential pressure across the test section increases due to wax depositing on the wall and thereby decreasing the effective pipe diameter. The effective diameter can be calculated from the pressure drop, which can be measured with differential pressure transducers/transmitters installed at the inlet and outlet of the test section.

Complications arise for the temperature characterization in the test section. Temperature measurements within the test section are usually not as satisfactory because the measuring device, such as a thermal couple, is likely to interfere with the oil flow field as well as having wax depositing on its surface, which greatly compromises its accuracy. Therefore, thermal couples are usually installed upstream of the test section inlet and downstream of the test section outlet to monitor the bulk temperature of the oil and the coolant.

It would be ideal to be able to measure the temperature at the inner pipe wall or deposit surface because this temperature represents the local temperature condition under which wax deposition occurs. However, the temperature close to a solid surface varies greatly within the boundary layer adjacent to the surface (an example is highlighted in Figure 5.3). The thickness of the boundary layer is typically less than 1 mm. Consequently, if the probe of a thermal couple is not placed exactly at the inner wall surface, the boundary layer of the flow can be interrupted, and the wall temperature might not be correctly measured.

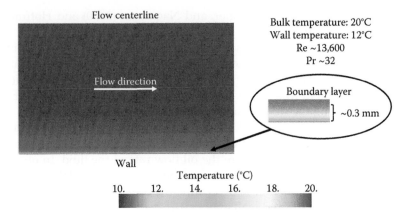

FIGURE 5.3
An example of the thickness of the boundary layer in normal turbulent pipe flow conditions.

5.3 Deposit Characterization

At the end of a flow-loop wax deposition experiment, a layer of wax deposit is formed on the inner pipe wall surface. The characterization of the deposit should be carefully carried out, as the measurement results often provide critical information for theoretical validation and model bench-marking. Two parameters are usually measured: the thickness of the deposit and the composition of the waxy components in the deposit. In this section we will discuss different methods for these investigations.

5.3.1 Measurement of the Deposit Thickness

For the past few decades, numerous techniques have been developed to measure the thickness of the wax deposit. Some of these techniques have been used in many experimental studies, while others still remain at development stage and are to be optimized before they can be used extensively. These techniques will be discussed in detail in this section.

5.3.1.1 Pressure-Drop Technique

The most common approach is the pressure-drop method because it can be carried with minimum interference to the oil flow throughout the course of a deposition experiment, because the pressure transducers are installed at the inlet and outlet of the test section. In this method, an empirical hydraulic correlation is often used to determine the effective diameter of the pipe and the deposit thickness. An example for single-phase oil flow is the Haaland correlation (Haaland, 1983), which was used in the deposition study of Hoffmann and Amundsen (2010). This correlation to calculate the effective pipe diameter is shown by

$$f_{Darcy} = \left\{ -1.8 \log_{10} \left[\left(\frac{\varepsilon_{pipe}}{3.7 d_{pipe}} \right)^{1.11} + \frac{6.9}{Re} \right] \right\}^{-2} \quad (4000 < Re < 10^8) \quad (5.1)$$

In this correlation, f_{Darcy} is the Darcy friction factor. It can be converted to the pressure drop across a horizontal pipe via the following (Wilkes, 2005):

$$\Delta P_{pipe} = \frac{8 f_{Darcy} \rho_{oil} L_{pipe} Q_{oil}^2}{\pi^2 d_{pipe}^5} \quad (5.2)$$

It can be seen from Equation 5.1 that the friction factor and the pressure drop depend on the roughness of the pipe, ε_{pipe}, and the Reynolds number of the flow, which are often associated with a degree of uncertainties.

Several caveats should be noted when using this correlation. First, as the wax deposit forms a layer of coverage on the pipe wall, the roughness of the deposit surface instead of that of the pipe should be used in Equation 5.1. However, the surface morphology of the deposit surface depends on multiple factors, including the composition of the waxy components as well as the operating conditions. Therefore, the roughness of the deposit varies from test to test. Based on the above Haaland correlation, the impact of the roughness is shown in Figure 5.4. The y-axis represents the ratio between the two

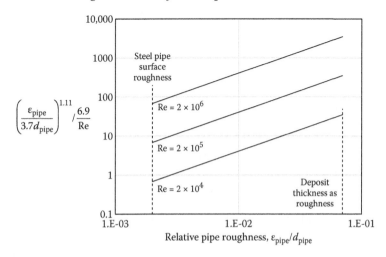

FIGURE 5.4
Contribution of friction pipe roughness on the friction factor at different values of Re.

terms inside the bracket on the right-hand side of Equation 5.1. This ratio reflects the impact of the roughness on the friction factor. It can be seen that such impact increases with increasing Reynolds number. For example, based on a 2-in. steel pipe for laboratory wax deposition experiments, the roughness of the bare pipe surface is approximately 50 μm. This value corresponds to a relative roughness ($\varepsilon_{pipe}/d_{pipe}$) of 2×10^{-3}. As shown in the dash line on the left of Figure 5.4, the ratio for a Reynolds number of 20,000 is less than 1, indicating that the friction factor (and thus the pressure drop) does not significantly depend on the roughness at this range. On the other extreme, the conclusion would be quite different if one prefers to use the thickness of the deposit as the pipeline roughness in this correlation, as was the approach in the study of Noville and Naveira (2012). In this case, the value of the roughness parameter can be up to a few millimeters. This value corresponds to a relative roughness of around 0.07 and is highlighted in the dash line on the right of Figure 5.4. The corresponding ratio for the same Reynolds number (20,000) is more than 50, indicating that the parameter of roughness has a significant influence on the pressure drop. In this case, as the thickness of the deposit is used as the roughness of the pipe, the pressure drop can be rather sensitive to the thickness of the deposit. Intuitively, the approach of using the deposit thickness as the pipeline roughness for the pressure drop–diameter correlation should be used only when the wax deposits are not expected to have complete coverage of the test section (for example, in some of the multiphase flow conditions where the oil does not have contact with part of the pipe wall).

For wax deposition experiments in single-phase oil flow where the deposits form a complete coverage in the test section, Hoffmann and Amundsen (2010) developed a technique that introduced short-term variations in the flow rate to estimate the roughness of the wax deposit surface at different operating conditions, and a range of 5–40 μm was reported.

The second parameter that is often associated with uncertainty is the viscosity of the oil, which is often used to determine the Reynolds number of the flow. Most of the available friction factor correlations are developed based on Newtonian fluid, where the viscosity used in determining the Reynolds number is independent of shear rates. However, the precipitated wax crystal in the oil flow can cause non-Newtonian behavior in the flow. For non-Newtonian flow, the viscosity depends on the shear rate in the flow field and thus is dependent on the oil flow rate. In addition, for a wax deposition experiment with a certain oil flow rate, the variation in the oil viscosity in the radial direction needs to be accounted for, as the shear rate greatly changes radially in the flow field. Therefore, methods such as multidimensional computational fluid dynamics are needed to incorporate the rheological characteristics of the non-Newtonian flow into the 2-D (axial and radial) momentum equation to evaluate the pressure drop across the test section in order to eventually determine the deposit thickness (Crochet, Davies,

& Walters, 1991). Given the complexity of flow in multidimensions, a rigorous empirical 1-D pressure-drop correlation does not appear to be feasible to describe the friction factor for non-Newtonian flow. Therefore, it is often preferred to compare the deposit thickness measured by the pressure drop with other methods (for example, weight measurements) to understand the range of the uncertainties.

For wax deposition experiments in multiphase flow, the fluid mechanics greatly depends on the flow patterns (stratified, annular, slug, or dispersed flow). In many cases, the flow is non-axisymmetric, and the deposit might not form uniformly on the pipe wall (for example, during gas–oil stratified flow conditions), and therefore, 1-D hydraulic correlations for multiphase flow would suffer too much uncertainty to be used to accurately measure the deposit thickness.

5.3.1.2 Weight Measurement Technique

For a test section that is equipped with a removable pipe, the weight of the deposit formed on this pipe can be obtained by measuring and comparing the weights of this pipe before and after the experiment. At the end of a wax deposition experiment, a thin layer of residual oil is often found on the deposit surface. In this case, nitrogen is commonly used to purge the pipe to remove the residual oil in order to establish correct weight measurements. Removing the residual oil is also important for the procedure of deposit sampling for further composition analysis.

To convert the weight of the deposit to the thickness of the deposit, the density of wax needs to be characterized. Hoffmann and Amundsen (2010) achieved this goal by scraping a deposit sample off the removable spool piece and placing the sample in a gas displacement pycnometer. This pycnometer uses gas as a displacement agent of which the equation of state is well established and calibrated. A small sample of the wax deposit (not the entire wax deposit in the removable pipe) with known weight is displaced with gas to determine its volume. The density of this small sample of wax deposit can thus be calculated. With this value of deposit density, the volume of the wax deposit in the entire removable pipe can then be calculated. The density of the deposit is often found to be slightly higher than that of the oil due to the shrinkage that occurs during its formation on the wall (Lee, 2008).

5.3.1.3 Heat Transfer Technique

The heat transfer technique utilizes the insulating effect of the wax deposit. The wax deposit accumulating on the pipe wall during the course of a wax deposition experiment acts as an additional insulation to the pipe, reducing the heat loss of the oil to the coolant. This reduction in the heat loss can be seen in the increase in the oil outlet temperature (Hernandez, 2002). The energy balance equations used to determine the thickness of the wax deposit are shown in Equations 5.3 through 5.5, and a schematic of an example in countercurrent flow is shown in Figure 5.5:

$$\Delta Q_{\text{thermal}} = \rho_{\text{oil}} C_p \left(T_{\text{oil,inlet}} - T_{\text{oil,outlet}} \right) = \pi U_{\text{overall}} d_{\text{outer}} L_{\text{pipe}} \Delta T_{\text{lm}} \qquad (5.3)$$

$$\Delta T_{\text{lm}} = \begin{cases} \dfrac{\left(T_{\text{oil,inlet}} - T_{\text{coolant,inlet}} \right) - \left(T_{\text{oil,outlet}} - T_{\text{coolant,outlet}} \right)}{\ln\left(\dfrac{T_{\text{oil,inlet}} - T_{\text{coolant,inlet}}}{T_{\text{oil,outlet}} - T_{\text{coolant,outlet}}} \right)} & \text{for cocurrent flow} \\[4em] \dfrac{\left(T_{\text{oil,inlet}} - T_{\text{coolant,outlet}} \right) - \left(T_{\text{oil,outlet}} - T_{\text{coolant,inlet}} \right)}{\ln\left(\dfrac{T_{\text{oil,inlet}} - T_{\text{coolant,outlet}}}{T_{\text{oil,outlet}} - T_{\text{coolant,inlet}}} \right)} & \text{for countercurrent flow} \end{cases}$$

$$(5.4)$$

$$\frac{1}{U_{\text{overall}}} = \frac{d_{\text{outer}}}{d_{\text{effective inner}}} \frac{1}{h_{\text{internal}}} + \frac{d_{\text{outer}}}{2k_{\text{pipe}}} \ln\left(\frac{d_{\text{outer}}}{d_{\text{inner}}} \right)$$

$$+ \frac{d_{\text{outer}}}{2k_{\text{deposit}}} \ln\left(\frac{d_{\text{inner}}}{d_{\text{effective inner}}} \right) + \frac{1}{h_{\text{coolant}}}$$

$$(5.5)$$

It can be seen from Equation 5.5 that the overall heat transfer resistance, $1/U_{\text{overall}}$, represents the summation of each thermal resistance in the radial direction (the oil phase, the deposit, the pipe wall, and the coolant). The thermal resistance of the oil phase, $\dfrac{d_{\text{outer}}}{d_{\text{effective inner}}} \dfrac{1}{h_{\text{oil}}}$, depends on the heat transfer

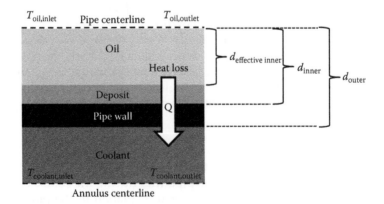

FIGURE 5.5
Schematic of radial heat transfer during the course of wax deposition.

coefficient in the oil phase, h_{oil}, which can be obtained by empirical correlations such as the Sieder and Tate correlation for turbulent flow:

$$h_{oil} = \frac{k_{oil}}{d_{effective\ inner}} Nu_{oil} = \frac{k_{oil}}{d_{effective\ inner}} \times 0.027\, Re^{0.8}\, Pr^{1/3} \left(\frac{\mu_{centerline}}{\mu_{T_{deposit\ interface}}} \right)^{0.14}$$

$$\begin{cases} 0.7 < Pr < 16,700 \\ Re > 10,000 \\ \dfrac{L_{pipe}}{d_{effective\ inner}} > 10 \end{cases} \tag{5.6}$$

It can be seen that a viscosity ratio, $\left(\dfrac{\mu_{T_{centerline}}}{\mu_{T_{deposit\ interface}}} \right)^{0.14}$, is used to account for the effect of nonuniform radial viscosity due to radial heat transfer. The viscosity of oil at the oil–deposit interface can be difficult to determine. In some cases, the viscosity ratio is assumed to be unity (Lund, 1998). The thermal resistance of the coolant phase, $\dfrac{1}{h_{coolant}}$, can be determined in a similar way by calculating the heat transfer coefficient of the coolant side using empirical correlations.

The other two thermal resistances are from the deposit layer and from the pipe wall, and their values depend on the parameters related to the size and the thermal properties of the deposit and the pipe wall. Among these parameters, the effective diameter of the oil flow is directly related to the thickness of the deposit, as shown in Equation 5.7. Consequently, by solving Equation 5.6 for the effective diameter of the oil flow, $d_{efective\ inner}$, the deposit thickness δ can be simply determined:

$$\delta_{deposit} = d_{inner} - d_{efective\ inner} \tag{5.7}$$

It should be noted that the heat transfer method suffers from a few disadvantages: First, for a laboratory-scale test section that is on the range from a few meters to tens of meters, the difference in the temperatures between the inlet and the outlet might not be significant. In this case, the uncertainty in the thermal couple measurements can greatly compromise the accuracy of the deposit thickness. In addition, the thermal conductivity of the deposit must be known a priori for this measurement. However, the deposit represents a mixture of liquid oil and wax crystals, and the wax crystal can have a thermal conductivity that is 50%–100% greater than that of the oil (Singh, 2000). Therefore, the thermal conductivity of the deposit greatly depends on the oil fraction of the deposit, which is not possible to determine during a wax deposition experiment. Even after a wax deposition experiment, challenges still exist where the oil fraction could change due to the change in temperature during the transfer of the deposit from the pipe to the measuring device. Consequently, the heat transfer method is not widely used to determine the deposit thickness.

5.3.1.4 Liquid Displacement–Level Detection Technique

The liquid displacement–level detection technique was developed by the University of Tulsa (Lund, 1998) and was used extensively during wax deposition experiments in the 2000s (Alana, 2003; Couto, 2004; Hernandez, 2002; Lund, 1998; Matzain, 1997). This technique determines the volume of the wax deposit by measuring the change in the volume of a removable pipe section before and after wax deposition. The measurement of the volume of the removable section after wax deposition is carried out by displacing water or other agents in the removable pipe section. The limitation of this technique comes when it is used to measure the thickness of soft or thin wax deposits, as this method is based on the subtraction of two large volumes (the volumes of the pipe before and after wax deposition). For thin deposits, the residual oil layer on top of the deposit can further increase the error of measurement. For soft deposits, part of the deposit can be removed during the measurement and thus is not accounted for.

5.3.1.5 Other Less Frequently Applied Techniques

5.3.1.5.1 Laser Technique

Hoffmann and Amundsen (2010) applied laser-based optical measurements of the wax deposit thickness in their single-phase and multiphase flow experiments. After a wax deposition experiment, a laser source was placed at the center of the pipe, which created a circular projection along the surface of the wax deposit, as shown in Figure 5.6. A camera was used to record this projected image and projections with the maximum light intensity corresponding to the radial location of the surface of wax deposit, based on a cylindrical coordinate with the center of the pipe as the origin. In this study, the measurements using the laser technique were compared with those

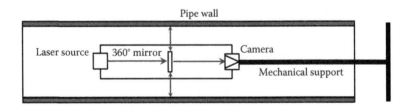

FIGURE 5.6
Schematic of the laser technique for deposit thickness measurement. (Reprinted with permission from Hoffmann, R., & Amundsen, L., *Energy & Fuels*, 24, 1069–1080. Copyright 2010 American Chemical Society.)

using the pressure-drop technique and weight technique, and the laser technique showed satisfactory agreement and consistency with other methods. However, due to interference with the flow field and device stability issues, this method can only be performed after the wax deposition experiment is completed and thus can only measure the final thickness of the wax deposit.

5.3.1.5.2 Acoustic Measurements

Acoustic measurements can be used to determine the thickness of the wax deposit by characterizing the change in the properties of the acoustic waves passing through the pipe. This method has the advantage that the measuring devices can be mounted on the external surface of the pipe wall, thereby circumventing the interference to the pipe flow. Consequently, this method can potentially provide continuous measurement of the deposit thickness during the course of wax deposition. Previous studies of using acoustic measurements include those by Esbensen et al. (1998), Lund (1998), and Halstensen, Arvoh, Amundsen, and Hoffmann (2013) in an attempt to determine the thickness of the wax deposit. However, it was reported that this method still has not been proven to have reliable consistency on the measurement, and it still remains to be further developed (Halstensen et al., 2013).

5.3.2 Composition Analysis of the Wax Deposit

During the course of wax deposition, the radial diffusion of the paraffin molecules causes their enrichment in the wax deposit. Consequently, the composition of the deposit provides an important parameter to characterize the degree of molecular diffusion for different waxy components. Singh et al. (2000) carried out one of the pioneering works to investigate the composition of the wax deposit with high-temperature gas chromatography (HTGC). It was found that there exists a threshold in the carbon number of the n-paraffins, termed as the critical carbon number (CCN). The n-paraffins with a carbon number above the CCN become enriched in the deposit during the course of wax deposition, while the n-paraffins with a carbon number below the CCN become depleted in the deposit. In this case, the change in the fraction of the waxy component n-C_iH_{2i+2} in the deposit ($F_i(t) - F_i(0)$) was calculated and is shown in Figure 5.7.

Hoffmann and Amundsen (2010) further showed that the CCN greatly depends on the oil composition as well as the operating conditions, as shown in Figure 5.7. The solubility of n-paraffins generally decreases as their carbon number increases. In other words, when the temperature decreases, heavier n-paraffins tend to precipitate earlier than lighter n-paraffins and thus more easily become enriched in the wax deposit. This difference in solubility of different n-paraffins explains the shift of the CCN shown in Figure 5.8.

The information of carbon number distribution in the wax deposits can provide critical information for wax deposition modeling, as will be shown in detail in Section 6.5.

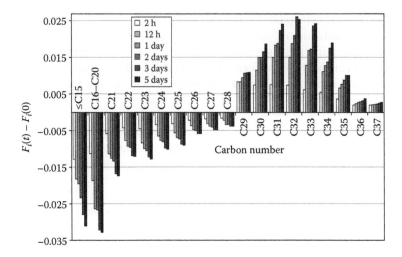

FIGURE 5.7
Enrichment of heavy waxy components during the course of wax deposition discovered in the study of Singh et al. (2000). (Singh, P., Venkatesan, R., Fogler, H. S., & Nagarajan, N.: Formation and aging of incipient thin film wax-oil gels. *AIChE Journal*. 2000. *46*. 1059–1074. Copyright Wiley-VCH Verlag GmbH & Co. KGaA. Reproduced with permission.)

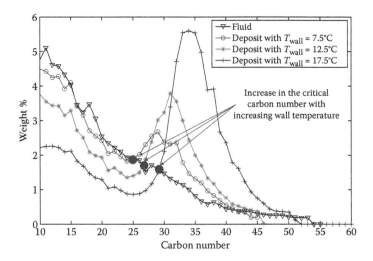

FIGURE 5.8
Increase in the CCN in the wax deposit for deposition experiments with increasing wall temperature. (Reprinted with permission from Hoffmann, R., & Amundsen, L., *Energy & Fuels*, 24, 1069–1080. Copyright 2010 American Chemical Society.)

5.4 Cold-Finger Wax Deposition Apparatus

Flow-loop deposition facilities are expensive to construct and operate. On the other hand, cold-finger wax deposition apparatus is less costly and requires a smaller volume of oil for testing. Many cold-finger apparatuses have frequently been used for the purpose of screening and selection of chemical additives (Jennings & Weispfennig, 2005).

A schematic example of two main components of a cold-finger deposition apparatus, i.e., the deposition cell and the circulation system, is shown in Figure 5.9. The deposition cell includes a metal cylinder that is usually referred to as the cold-finger probe. Conduits are made inside the probe for cold fluids (glycol or water) to circulate in order to maintain the probe at low temperatures, so that wax will deposit on its outer surface. During the course of a wax deposition experiment, the probe is placed inside an oil container. The container is often equipped with a stirrer to introduce shear to the system. The flow field of the cold finger is mainly Taylor–Couette type (azimuthal flow between two cylinders) with a certain degree of axial circulation. This type of flow is quite different from normal pipe flow at the field (Senra, 2009). Similar to deposition experiments in flow loop, residual oil can adhere to the deposit formed on the cold finger. In order to remove the residual oil, the cold finger with the deposit on the probe is dipped in methyl ethyl ketone to wash away the residual oil from the surface. To increase the efficiency of the experiment, several deposition cells can be placed inside a large heating bath so that multiple cold-finger experiments with different probe (cold finger) temperature can be carried out at the same time (Couto, 2004). The circulation system includes a heating system and

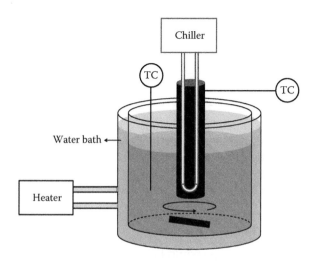

FIGURE 5.9
Schematic of the cold-finger wax deposition apparatus.

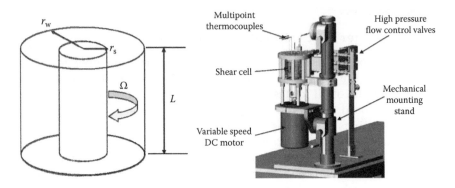

FIGURE 5.10

Schematic of the organic solid deposition cell in the study of Akbarzadeh and Zougari (2008). (Reprinted with permission from Akbarzadeh, K., & Zougari, M., *Industrial and Engineering Chemistry Research*, 47, 953–963. Copyright 2008 American Chemical Society.)

a cooling system. Temperature measurements are often used to maintain the thermal gradient between the oil and the cold-finger probe. A variation of the cold-finger deposition apparatus is the "organic solids deposition cell" where a rotating probe is used instead of a stirrer (Akbarzadeh & Zougari, 2008). In this case, the probe, often referred to as a "spindle," extends to the bottom of the oil container, as shown in Figure 5.10. One of the disadvantages of the nonflow-loop devices is "wax depletion" of the oil, which refers to the reduced deposition tendency during the course of deposition due to the decrease in the bulk concentration of the waxy components. The impact of wax depletion can be significant if the oil being tested has low wax content. In addition, because the flow field in these devices is significantly different from that of the pipe, it has not been used extensively for wax deposition modeling applications.

5.5 Carrying Out Flow-Loop Wax Deposition Experiments

In this section we would like to provide a brief introduction of a general wax deposition experimental procedure. Detailed descriptions of setting up a flow-loop wax deposition experiment, several examples can be found in the studies of Matzain (1997), Lund (1998), Singh et al. (2000), Hernandez (2002), Alana (2003), Couto (2004) and Hoffmann and Amundsen (2010).

5.5.1 Setup of Experimental Apparatus

Prior to testing, the flow loop should be tested for leakage. The flow meters, the pressure transducer, and the thermal couples should also be calibrated to ensure that the measurements are reliable.

5.5.2 Oil Characterization

The properties of oil are of great importance to comprehend the wax deposition testing results. Consequently, the oil should be well characterized. The composition of the oil (especially the n-paraffin distribution) should be measured, and wax precipitation tests should be carried out (detailed wax thermodynamic characterizations and modeling are introduced in Chapters 2 and 3, respectively). The viscosity of the oil should be measured at different shear rates and temperatures to assess the potential non-Newtonian behavior that might be caused by the precipitated wax crystals. These pre-deposition characterizations provide essential information on wax–oil thermodynamics and are an integral part to benchmarking wax deposition models with the deposition experiments.

5.5.3 Wax Deposition Tests

Before the deposition experiment, the oil is heated to a temperature well above the WAT and circulated in the entire flow loop for an extended period of time to remove wax deposits in the test section from previous experiments. The temperatures of the oil and the coolant are to be decreased to the desired operating conditions. The decrease in the oil and the coolant temperature should be carefully designed in order to avoid the formation of wax deposits before the oil temperature reaches the target value. During the course of a wax deposition, the deposit growth is monitored by measuring the pressure drop across the test section.

5.5.4 Measurement of Deposit Thickness and Characterization of Wax Deposits

At the end of the wax deposition experiment, the flow is stopped, and the wax deposit ceases to grow. Most of the flow-loop apparatuses consist of a removable section where further deposit characterization can be made. The surface of the deposit is often covered with a residual oil layer, which can interfere with the deposit characterizations, and N_2 is often used to flush to the test section to remove this residual oil layer.

The deposit characterizations include thickness and compositional analysis. The thickness of the deposit can be measured with a caliper on the deposit surface. In addition, the weight of the wax deposits can also be measured by comparing the weights of the removable section before and after the experiment. The three methods of thickness measurements (caliper, weight, and pressure drop) can be used in combination to identify the uncertainties in the results. The composition of the wax deposits can be obtained via the HTGC. Both the characterization of the amount and the composition of the deposit provide important information for benchmarking wax deposition models.

5.6 Summary

In Chapter 2, we discussed different experimental techniques to study *wax precipitation*. In this chapter, we introduced several typical *wax deposition* apparatuses and a general procedure to carry out flow-loop wax deposition experiments. The information as well as the referenced studies in these two chapters provide a comprehensive description on the various tools for those who look forward to developing a laboratory facility or extending an existing laboratory to analyze wax precipitation/deposition. In Chapter 6, we will discuss the results of a series of wax deposition experiments and reveal some of the most influential parameters governing the physics of wax deposition through an in-depth theoretical analysis on the experimental results.

6

Applying Wax Deposition Models to Flow Loop Experiments

6.1 Introduction

The goal of wax deposition modeling is to achieve rigorous and reliable predictions on the severity of wax deposition for the pipelines in the field. To achieve this goal, developing a comprehensive understanding of the performance and limitations of wax deposition models is essential. While field data would be the ideal candidate to directly test the wax deposition models, it is usually not available nor as well controlled and monitored.

Contrarily, laboratory wax deposition experiments are able to provide unique opportunities to benchmark wax deposition models due to ease of experimental design and characterization. The results from laboratory experiments can often help us develop educated guesses for the pipelines in the field. The process of applying wax deposition models to well-characterized flow-loop experiments helps one select the most appropriate deposition model, to understand the governing physics for the deposition phenomenon, and to identify the optimum configuration of the adjustable parameters in a wax deposition model. In this chapter, the process of conducting wax deposition experiments and the analyses of these experiments with wax deposition models will be applied.

6.2 Uncertainties in Wax Deposition Modeling

A wax deposition model provides the wax deposit thickness as a function of time and, in some cases, the evolution of the composition of the deposit during the course of wax deposition. Before using a wax deposition model to run a number of simulations that predict the deposit thickness, it is beneficial to understand the uncertainties associated with wax deposition modeling. Consequently, those uncertainties will now be identified and discussed.

6.2.1 Characterizing the Wax Precipitation Curves

The wax precipitating curve is one of the most important input variables for wax deposition modeling, as it contains information as to how much wax remains in the oil at various temperature/pressure conditions, assuming thermodynamic equilibrium. However, as previously discussed in detail in Chapter 2, characterization of the wax precipitation curve is not always perfect. For example, measurements from differential scanning calorimetry (DSC) rely on a constant averaged value of the heat of crystallization for all the waxy components. Because the waxy components with different carbon numbers exhibit different precipitation characteristics with temperature, the averaged value of the heat of crystallization might not be constant during the cooling of the sample. In fact, as the values of the heat of crystallization vary between 100 and 300 J/g (Hansen, Larsen, Pedersen, Nielsen, & Rønningsen, 1991), a middle point of 200 J/g is frequently selected for many academic and industrial purposes (regardless of the actual composition of the oil sample) as a simplification.

As an alternative to experimental measurements, thermodynamic models are often used to predict the wax precipitation curve based on a known n-paraffin composition of the oil (as were discussed in detail in Chapter 3). However, many of the models assume ideal solutions for the solid phase. This assumption is generally not strictly correct. Even if one tunes the model to match the wax appearance temperature (WAT; which represents the first data point in the wax precipitation curve), there is no guarantee that the entire wax precipitation curve (the amount of wax precipitation at temperatures except the WAT) is predicted accurately.

6.2.2 Empirical Correlations for Transport Phenomena

In addition to the wax precipitation curve, wax deposition models use transport calculations to determine the degree of wax deposition on the pipe wall during flow conditions. Most of the wax deposition models rely heavily on empirical correlations for heat and mass transfer. For example, the heat transfer coefficient for the oil in the pipe as well as the coolant in the annulus are often obtained from the correlations of Sieder and Tate (1936) and Monrad and Pelton (1942), respectively. These empirical correlations were developed in the mid-1900s and are known to have a certain level of experimental uncertainties due to the limitation in the measuring technology at the time.

Additionally, the use of empirical correlations for mass transfer is required for the mass diffusivity for wax (also known as the wax diffusion coefficient). This parameter is often determined by the correlations from the studies of either Hayduk and Minhas (1982) or Wilke and Chang (1955). These correlations were developed based on binary dilute solutions, and uncertainties can arise when applying them to crude oils with polydispersed waxy components and with much higher concentrations.

6.2.3 Uncertainties in Experimental Measurements

One of the main outputs of a wax deposition model is the thickness of the wax deposit, which is often compared with the values measured in an experiment. The most frequently used experimental method for the deposit thickness is through measuring the pressure drop in the pipe. With empirical hydrodynamic correlations such as the one by Haaland (1983), the pressure drop in the pipe can be used to determine the thickness of the deposit. These correlations often greatly depend on the viscosity of the oil and the roughness of the wall. The viscosity can be greatly complicated by the non-Newtonian behavior of the oil, which is not accounted for in most of the correlations that have been used to determine the deposit thickness. As discussed in detail in Chapter 5, the roughness of the wall becomes complicated as the wall begins to become covered with wax deposit during the course of wax deposition. The exact roughness of the deposit surface has not been rigorously determined. While some studies observed that the roughness of the deposit surface is around 5–40 μm (Hoffmann & Amundsen, 2010), others believe that the thickness of the deposit should be the roughness of the pipe (Noville & Naveira, 2012).

6.2.4 Appropriate Methodology for Wax Deposition Benchmarking

The above uncertainties might have cast a pessimistic view on the use of flow-loop experiments to benchmark wax deposition models. However, this is generally true for most model validations: *Fitting the numerical predictions with only few sets of experimental data and without any sensitivity analysis should probably be taken with a grain of salt.*

Unfortunately for most industrial applications, time and cost can become significant constraints for a project. In many circumstances, the approach of tuning a wax deposition model using just one or two experiments and lump all of the uncertainties into one or two adjustable parameters becomes a convenient choice.

In fact, many industrial practices have not even applied laboratory benchmarking for the deposition models. However, in order to evaluate and compare the predictions from different wax deposition models, an understanding of the governing physics and the identification of the key parameters for wax deposition is necessary. The approach to fit a wax deposition model with just one or two experimental results might not yield in-depth conclusions that would be applicable for other cases with different oils and with different operating conditions. Consequently, it would be beneficial to apply wax deposition models to a series of flow-loop experiments with a wide range of varying operating conditions to establish a systematic understanding as to how the wax models can capture the physics represented by the experimental results. For many industrial practices, there might not always be sufficient resources to carry out such a task. Consequently, examples of comprehensive model validations will be discussed in detail in Sections 6.3 and 6.4.

A wax deposition experiment often provides two main results: the thickness of the deposit as well as the composition of the wax deposit as a function of time. The thickness of the deposit is measured in nearly all of the wax deposition experiments, and it is often selected as the variable to benchmark wax deposition models. The composition of the deposit provides the most valuable insights as to the diffusion of different waxy components during the course of deposition. Unfortunately, it is not always measured and analyzed. The theoretical analysis on these two types of results will be discussed extensively in the rest of this chapter.

6.3 Applying Wax Deposition Models with Deposit Thickness

6.3.1 Selecting Wax Deposition Experiments

This chapter will focus on a comprehensive series of flow-loop experiments carried out in the Herøya Research Center of Statoil ASA in Porsgrunn, Norway (Hoffmann & Amundsen, 2010; Huang, Lu, Hoffmann, Amundsen, & Fogler, 2011; Lu et al., 2012). This experimental facility, the oil used for the experiments, as well as the testing conditions will now be introduced followed by the discussion of the experimental results.

6.3.1.1 Experimental Facility

The flow loop is equipped with an oil storage tank of 4 m³ and a test section with a length of 5.5 m and an inner diameter of 5.3 cm. The flow loop is equipped with a pipe-in-pipe test section. The testing fluid flows through the center of the test section, and the coolant (water) flows through the annulus shell. A schematic of the flow-loop facility is shown in Figure 6.1.

6.3.1.2 Test Oil

The oil used for the experiments is a waxy gas condensate from the North Sea with a density of 809 kg/m³ and a WAT of 20°C measured by DSC. Measurements on the condensate using acetone precipitation (UOP 46–64 method) revealed a wax content of approximately 4.5 wt%. The viscosity of the oil was characterized at different shear rates using a Physica MCR 301 rheometer, as shown in Figure 6.2.

The wax precipitation curve was measured by several methods, as shown in Figure 6.3. The details of these methods were previously introduced in Section 2.2. The method of characterization using centrifugation and HTGC was previously reported by Han, Huang, Senra, Hoffmann, and Fogler (2010). Another method utilizes the DSC based on an averaged value of the heat of crystallization of 200 J/g. Considering that the actual heat

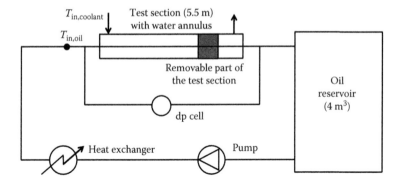

FIGURE 6.1
Schematic of the flow loop in Statoil's Herøya Research Center. (From Hoffmann, R., & Amundsen, L. *Energy & Fuels*, 24, 1069–1080, 2010; Huang, Z. et al., *Energy & Fuels*, 25, 5180–5188, 2011.)

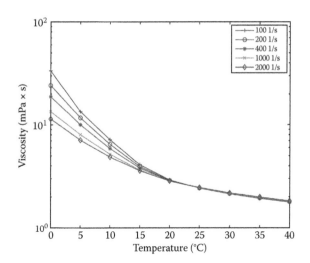

FIGURE 6.2
Viscosity curves of the North Sea condensate at different shear rates. (Reprinted with permission from Hoffmann, R., & Amundsen, L. *Energy & Fuels*, 24, 1069–1080. Copyright 2010 American Chemical Society.)

of crystallization of the waxy components often lies between 300 and 100 J/g (Hansen et al., 1991), the methods of centrifugation + HTGC and DSC appear to be consistent with one another.

6.3.1.3 Experimental Conditions

Three series of wax deposition experiments were carried out: one series with different oil temperatures, another with coolant temperatures, and the third

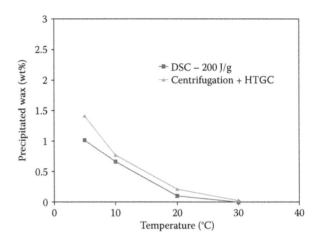

FIGURE 6.3

The precipitation curve of the North Sea condensate characterized with different methods. (Reprinted with permission from Han, S., Huang, Z., Senra, M., Hoffmann, R., & Fogler, H. S. *Energy & Fuels*, 24, 1753–1761. Copyright 2010 American Chemical Society.)

one with different oil flow rates. The detailed configurations of these series of experiments are summarized in Table 6.1.

The thickness of the wax deposit was determined by three methods: the pressure-drop technique, the weight technique, and the laser technique. The pressure-drop measurement was nonintrusive and was carried out continuously during the course of wax deposition, while intrusive methods like

TABLE 6.1

Summary of the Series of Wax Deposition Experiments Carried Out in Statoil's Herøya Research Center

	Oil (Inlet) Temperature (°C)	Coolant (Inlet) Temperature (°C)	Oil Flow Rate (m³/h)	Coolant Flow Rate (m³/h)
Varying oil temperature	15	5	20	5
	20			
	25			
	35			
Varying coolant temperature	20	5	5	
		10		
		15		
Varying oil flow rate	20	10	5	
			10	
			15	
			21	
			25	

weight and laser measurements were carried out at the end of the experiment. A comparison of these three methods is highlighted in Figure 6.4, indicating the consistency among these measurement techniques. The reproducibility of the experiments is also confirmed by repeating one of the experimental conditions listed in Table 6.1 four times (oil temperature: 20°C; coolant temperature: 10°C; oil flow rate: 5 m³/h; and coolant flow rate: 5 m³/h), as shown in Figure 6.5.

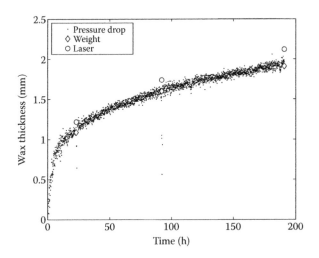

FIGURE 6.4
Comparison of the deposit thickness measured by different methods. (Reprinted with permission from Hoffmann, R., & Amundsen, L. *Energy & Fuels*, 24, 1069–1080. Copyright 2010 American Chemical Society.)

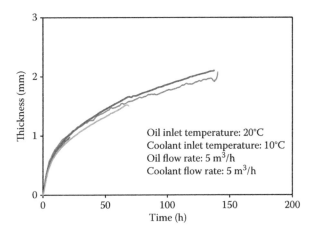

FIGURE 6.5
Reproducibility of the wax deposition experiments.

6.3.2 Summary of Model Performance

A wax deposition model called the "Michigan Wax Predictor" (MWP) was used to model the series of experiments shown in Table 6.1. The transport equations for the MWP have been discussed previously in Chapter 4. Instead of tuning the wax model to match each experiment, we chose to eliminate any tuning parameter in the wax model and investigate if the model can predict the trends shown in the experiments. The reason is that it is expected that the experimental uncertainties could possibly impact the deposit thickness in similar directions. Therefore, by looking at the experimental trends

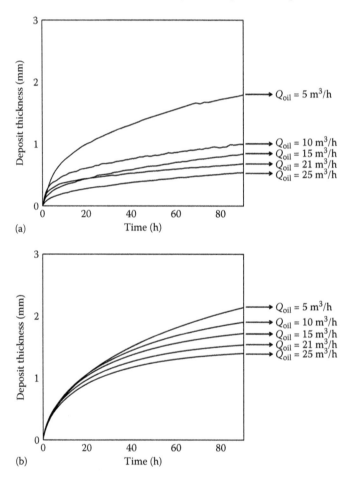

(a)

(b)

FIGURE 6.6

Comparison of deposit thickness as a function of oil flow rate, Q_{oil}, between (a) the experimental results and (b) the model prediction by the MWP. The oil temperature, T_{oil}, was maintained constant at 20°C, and the coolant temperature $T_{coolant}$, was maintained constant at 10°C. (Reprinted with permission from Lu, Y., Huang, Z., Hoffmann, R., Amundsen, L., Fogler, H. S., & Sheng, Z. *Energy & Fuels*, 26, 4091–4097. Copyright 2012 American Chemical Society.)

shown from a group of experiments (rather than each individual experiment), the impact of those uncertainties can be reduced. In this case, the independent heat and mass transfer submodel was used (Huang, Lee, Senra, & Fogler, 2011; Lee, 2008; Singh, Venkatesan, Fogler, & Nagarajan, 2000). The model predictions are compared with the experimental measurements, as shown from Figures 6.6 through 6.8.

It is seen that the MWP was able to reproduce all the changes in the deposit thickness when the operating conditions (oil flow rate, oil temperature, and/or coolant temperatures) change. The trends seen in the experiment and the model predictions are summarized in Table 6.2.

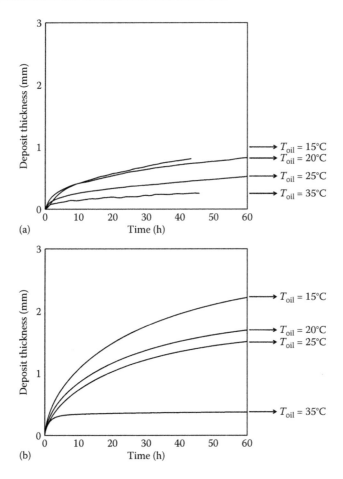

FIGURE 6.7
Comparison of deposit thickness as a function of oil temperature, T_{oil}, between (a) the experimental results and (b) the model prediction by the MWP. The oil flow rate, Q_{oil}, was maintained constant at 20 m³/h, and the coolant temperature $T_{coolant}$ was maintained constant at 5°C. (Reprinted with permission from Huang, Z., Lu, Y., Hoffmann, R., Amundsen, L., & Fogler, H. S. *Energy & Fuels*, 25, 5180–5188. Copyright 2011 American Chemical Society.)

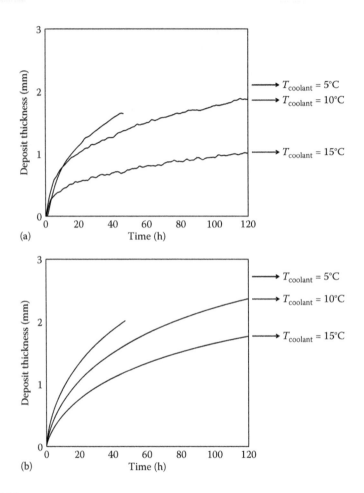

(a)

(b)

FIGURE 6.8

Comparison of deposit thickness as a function of coolant temperature, $T_{coolant}$, between (a) the experimental results and (b) the model prediction by the MWP. The oil flow rate, Q_{oil}, was maintained constant at 5 m³/h, and the oil temperature T_{oil}, was maintained constant at 20°C (Reprinted with permission from Huang, Z., Lu, Y., Hoffmann, R., Amundsen, L., & Fogler, H. S. *Energy & Fuels, 25,* 5180–5188. Copyright 2011 American Chemical Society.)

TABLE 6.2

Summary of the Experimental Trends and That Predicted by the MWP

Operating Conditions	Experimental Trend	MWP
Increased flow rate	Decreased thickness	Decreased thickness
Increased oil temperature	Decreased thickness	Decreased thickness
Increased coolant temperature	Decreased thickness	Decreased thickness

It should be noted that the MWP was not used to exactly match the value of the thickness of the wax deposit for each individual experiment. It is believed that it is neither practical nor would it provide significant insight to reveal the physics of wax deposition by tweaking one or two adjustable parameters to match all the experimental data, given the degree of uncertainties associated with wax deposition modeling that has been discussed in Section 6.2.3. However, the MWP was able to predict all the experimental trends, and the importance of this achievement cannot be overstated: First, the change in the oil flow rate and in the oil temperature is essential for the model scale-up for field pipelines, as the oil flow rate in these laboratory experiments ranges from 5 to 25 m³/h; while the flow rate in the field pipelines often ranges from 10,000 to 30,000 barrels per day (bpd) (i.e., 66–200 m³/h). While the oil temperature remains virtually constant in the 5.5 m long test section for the laboratory flow-loop experiments, the actual oil temperature in the field pipelines can change from a value above 70°C down to close to the ocean floor temperature of about 4°C. Consequently, being able to capture the correct direction of change in the wax deposit thickness during the change in these operating conditions is critical to the development of reliable predictions to the field. To the authors' knowledge, we know of no other wax deposition model that has been reported to have this success. Because of this success, more in-depth heat and mass transfer analysis will be presented in Section 6.4 to help reveal the most influential parameters for wax deposition based on the studies of Huang, Lu et al. (2011) and Lu et al. (2012).

6.4 Heat and Mass Transfer Analysis of the Wax Deposition Experiments

In this section, the answers to an important question will be analyzed: What are the elements that cause the changes in the wax deposit thickness when the operating conditions change? In other words, given the experimental trends presented previously in Table 6.2, *does this trend hold for other oils in completely different operating conditions?* Consequently, it would be beneficial to first review the trends shown in several other previous experimental studies.

6.4.1 Previous Wax Deposition Experiments on the Effect of Temperature

Because wax deposition is a result of radial temperature difference, this difference has been referred to as the "thermal driving force" for wax deposition in several experimental studies (Bidmus & Mehrotra, 2009; Creek, Lund, Brill, & Volk, 1999; Singh et al., 2000). In these studies, the thermal driving force often refers to the temperature difference between the bulk oil and the coolant. Wax deposition studies were often carried out with different thermal driving forces. In the study by Jennings and Weispfennig (2005), a crude oil from the Gulf of Mexico was used in a cold-finger wax deposition apparatus. It was found that experiments with higher cold-finger temperatures (smaller thermal driving force) yielded smaller amounts of deposited wax. Creek et al. (1999) found similar results with a flow loop in which the deposit thickness decreased when the coolant temperature increased (decreasing thermal driving force). Consequently, it was often believed that the deposit thickness would decrease when the thermal driving force decreases.

However, exceptions were, in fact, observed in other studies. In the cold-finger study by Paso and Fogler (2004), it was observed that lower oil temperatures (a smaller thermal driving force) lead to thicker wax deposits. In the study by Bidmus and Mehrotra (2009) using a mini flow loop, different trends were observed: When the oil temperature is greater than the WAT (referred to as "hot flow"), the amount of wax deposit decreases with increasing thermal driving force. When the oil temperature is below the WAT (referred to as "cold flow"), the amount of wax deposit decreases with decreasing thermal driving force.

Given the many different experimental observations reported, it would first appear that there is no established conclusion on the behavior of wax deposition with varying temperatures. Indeed, with different oils and different operating conditions, the observation in one set of experiments from one oil sample might not necessarily be appropriate for another. However, *the fundamental heat and mass transfer equations for these experiments are the same or rather similar*, giving us some hope that the transport theories can be utilized to identify a simple rule that can explain these different observations.

6.4.2 Theoretical Analysis

In this section, an in-depth mass transfer analysis will be made to identify the underlying physics that governs the behavior of wax deposition. To achieve this goal, one of the classic approaches in transport phenomena—dimensionless analysis—will first be applied.

6.4.2.1 Dedimensionalizing the Transport Equations

A schematic of the mass transfer characteristics in normal pipe flow conditions is shown in Figure 6.9. The equation used in the MWP to describe this process is shown in Equation 6.1 with the following boundary conditions shown in Equation 6.2:

$$V_z \frac{\partial C}{\partial z} = \frac{1}{r} \frac{\partial}{\partial r} \left[r(\varepsilon_{mass} + D_{wax}) \frac{\partial C}{\partial r} \right] \tag{6.1}$$

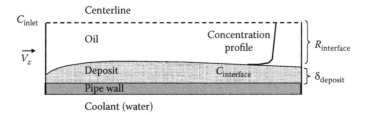

FIGURE 6.9
Schematic of the wax concentration profile at normal pipe flow conditions.

$$\begin{cases} \text{at } z = 0,\ C = C_{\text{inlet}} \\[2ex] \text{at } r = 0,\ \dfrac{\partial C}{\partial r} = 0 \\[2ex] \text{at } r = R_{\text{interface}},\ C = C_{\text{interface}} \end{cases} \tag{6.2}$$

where C represents the concentration of the dissolved wax molecules in the oil; V_z represents the axial velocity field of the oil flow in the pipe; $\varepsilon_{\text{mass}}$ is the eddy mass diffusivity when the flow is turbulent; D_{wax} is the mass diffusivity of wax in oil (also known as the wax diffusion coefficient) of the wax molecule in oil; and r and z are the two coordinates considered in the system. It can be seen that Equation 6.1 represents a typical 2-D Graetz problem with axial advection and radial diffusion.

With the dimensionless ratios shown in Equation 6.3, the mass balance equation and its boundary conditions can be transformed into Equations 6.4 and 6.5:

$$\left(\theta = \frac{C - C_{\text{interface}}}{C_{\text{inlet}} - C_{\text{interface}}} \right),\left(v = \frac{V}{U} \right),\left(\lambda = \frac{z}{L} \right),\left(\eta = \frac{r}{r_{\text{d}}} \right),\left(Gz = \frac{d_{\text{d}}^2 U}{L(\varepsilon_{\text{mass}} + D_{\text{wo}})} \right) \tag{6.3}$$

$$v\frac{\partial \theta}{\partial \lambda} = \frac{1}{\eta}\frac{\partial}{\partial \eta}\left[\frac{4}{Gz}\eta\frac{\partial \theta}{\partial \eta} \right] \tag{6.4}$$

$$\begin{cases} \text{at } \lambda = 0,\ \theta = 1 \\[2ex] \text{at } \eta = 0,\ \dfrac{\partial \theta}{\partial \eta} = 0 \\[2ex] \text{at } \eta = 1,\ \theta = 0 \end{cases} \tag{6.5}$$

It can be seen that the dimensionless concentration, θ, is independent of C_{inlet} and $C_{interface}$. This dimensionless form gives us the benefit to describe the critical parameter for wax deposition—the radial mass flux of wax molecules (see J_A in Equation 4.10 in Chapter 4), which can now be rearranged, as shown in Equation 6.6:

$$J_A = D_{wax,interface} \left.\frac{dC}{dr}\right|_{interface} = D_{wax,interface} \frac{\partial C}{\partial \theta} \cdot \left(\left.\frac{\partial \theta}{\partial r}\right|_{interface}\right)$$

$$= \frac{D_{wax,interface}(C_{inlet} - C_{interface})}{r_d} \left(\left.\frac{\partial \theta}{\partial \eta}\right|_{interface}\right) = J_{wax} \left(\left.\frac{\partial \theta}{\partial \eta}\right|_{interface}\right) \quad (6.6)$$

$$J_{wax} = \frac{D_{wax,interface}(C_{inlet} - C_{interface})}{r_d} \quad (6.7)$$

In this case, the radial mass flux J_A can be seen as a multiplication of two parameters: The first one, noted as J_{wax}, appears to have the same unit of a mass flux and is called the "characteristic mass flux" for wax deposition, while the second parameter, the dimensionless concentration gradient, $\left(\left.\frac{\partial \theta}{\partial \eta}\right|_{interface}\right)$, depends mainly on the oil flow rate.

6.4.2.2 Characteristic Mass Flux of Wax Deposition

The parameter of the characteristic mass flux of wax deposition, J_{wax}, has significant physical representation: By analyzing this parameter, the impact of temperature on wax deposition can be almost revealed completely.

From Equation 6.7, it can be seen that J_{wax} consists of two terms that are dependent on temperature: the concentration difference ($C_{inlet} - C_{interface}$) and the diffusivity of wax in oil at the interface, $D_{wax,interface}$. The first term contains information about *thermodynamic* equilibrium, and the second one represents the *transport* characteristics of waxy components in oil. Consequently, this arrangement in Equation 6.6 gives us the advantage to independently quantify the influence of temperature on both of these two aspects. First, by assuming thermodynamic equilibrium for the wax concentrations C_{inlet} and $C_{interface}$, the characteristic mass flux for wax deposition can be approximated by the following equation:

$$J_{wax} = \frac{D_{wax,interface}(C_{inlet} - C_{interface})}{R_{interface}} \approx \frac{D_{wax,interface}[C_{inlet}(eq) - C_{interface}(eq)]}{R_{interface}} \quad (6.8)$$

In addition, the effect of the temperature on the diffusivity can be estimated using the Hayduk–Minhas correlation (Hayduk & Minhas, 1982), as shown by

$$D_{wax,interface} = 13.3 \times 10^{-12} \frac{T_{interface}^{1.47} \mu_{deposit\ interface}^{(10.2/V_A - 0.791)}}{V_A^{0.71}} \text{m}^2/\text{s} \tag{6.9}$$

Difficulties arise when one tries to evaluate $D_{wax,interface}$ and $C_{interface}(eq)$ because they depend on the interface temperature, $T_{interface}$, which in turn greatly depends on the predicted value of an existing deposit thickness. Therefore, one needs to know the exact thickness of the deposit in order to accurately determine both $C_{interface}(eq)$ and $D_{wax,interface}$. This hurdle can be resolved by evaluating the values of $D_{wax,interface}$ and $C_{interface}(eq)$ initially (at $t = 0$) where deposition has not yet occurred. In this case, $C_{interface}(eq)$ simply reduces to $C_{wall}(eq)$, $D_{wax,interface}$ reduces to $D_{wax,wall}$, and $R_{interface}$ becomes R_{pipe}, as shown in the following equation:

$$J_{wax}(t = 0) = \frac{D_{wax,wall}\left[C_{inlet}(eq) - C_{wall}(eq)\right]}{R_{pipe}} \tag{6.10}$$

Several parameters in Equation 6.10 depend on the temperature at the pipe wall, which can be determined based on the energy balance equation in the radial direction, as shown in Equation 6.11, along with well-established correlations for the heat transfer coefficients. For example, the correlation of Dittus and Boelter (1985) can be used to determine the heat transfer coefficient of the oil, h_{oil}, and the correlation of Monrad and Pelton (1942) can be used to determine the heat transfer coefficient of the coolant, $h_{coolant}$:

$$h_{oil}(T_{oil} - T_{wall}) = h_{coolant}(T_{wall} - T_{coolant}) \tag{6.11}$$

Once we have T_{wall}, we can accurately determine $C_{wall}(eq)$ from the wax solubility curve and $D_{wax,wall}$ from the Hayduk–Minhas correlation (Hayduk & Minhas, 1982). It should be noted that in most of the lab-scale flow-loop studies, no significant axial variation in the bulk concentration was observed for the parameters due to the short length of the pipe in the flow-loop test section. Therefore, $C_{inlet}(eq)$ is referred to as $C_{oil}(eq)$ for subsequent quantitative analyses.

6.4.2.3 Mass Transfer Driving Force

Within the characteristic mass flux for wax deposition, the difference in the concentrations between the bulk and the wall, $[C_{oil}(eq) - C_{wall}(eq)]$, is referred to as the *mass transfer driving force* in comparison of the thermal driving force, $[T_{oil} - T_{coolant}]$. In Section 6.4.3, both the mass transfer driving force and the thermal driving force will be used in our analysis on wax deposition behavior to identify which parameter of these two provides more insights to explain the trends observed in the experiments.

6.4.3 Effect of the Operating Temperatures

6.4.3.1 Analysis on the North Sea Condensate

Table 6.3 summarizes the flux term, J_{wax}, and its associated parameters for the experiments where the oil temperature, T_{oil}, was varied and the coolant temperature, $T_{coolant}$, was kept constant. In these experiments, the deposition thickness decreased as the oil temperature increases, as shown previously in Figure 6.7.

One observes that the increase in the oil temperature, T_{oil}, leads to an increase in the wall temperature, T_{wall}. Such increase in turn had several effects on the characteristic mass flux, J_{wax}: First, in terms of transport characteristics, it increased the diffusivity of wax in oil at the wall, $D_{wax,wall}$, which tended to increase the mass flux. Second, in terms of thermodynamic properties, the change in the oil and the wall temperatures caused an increase in the wax solubility, $[C_{oil}(eq) - C_{wall}(eq)]$. A close examination of the changes in these two terms reveals that the increase in $C_{oil}(eq)$ was from 1.09% to 1.44% (an increase of 35%). Such a degree of increase was less than that in $C_{wall}(eq)$, which increased from 0.65% to 1.26% (an increase of 190%). Meanwhile, the change in T_{oil} from 15.3°C up to 35.4°C (a change of 20.1°C) exceeded the change in T_{wall} from 9.5°C to 20.5°C (a change of 11°C). To summarize, although the bulk temperature increased more than the wall temperature, the wax solubility at the bulk increased much less than that at the wall. Consequently, while the thermal driving force $[T_{oil} - T_{coolant}]$ increased when the oil temperature increased, the mass transfer driving force $[C_{oil}(eq) - C_{wall}(eq)]$ actually decreased with increasing oil temperature.

Consider the variation in the other parameter: In this case, the coolant temperature, $T_{coolant}$, was varied, and the oil temperature, T_{oil}, was kept constant. For these experiments, it was found that the deposit thickness decreased as the coolant temperature increased (as shown previously in Figure 6.8). The

TABLE 6.3

Comparison of the Parameters Involved in the Characteristic Mass Flux of Wax Deposition, J_{wax}, among the Deposition Experiments with Different T_{oil} while Q_{oil} and $T_{coolant}$ Remained Constant

Parameters	Value			
T_{oil} (°C)	15.3	20.3	25.3	35.4
$T_{coolant}$ (°C)		5.0		
Q_{oil} (m³/h)		20.0		
T_{wall} (°C)	9.5	12.0	14.7	20.5
$D_{wax,wall} \times 10^{10}$ (m²/s)	2.11	2.49	2.89	3.64
$C_{oil}(eq)$ (wt%)	1.09	1.26	1.34	1.44
$C_{wall}(eq)$ (wt%)	0.65	0.89	1.07	1.26
$C_{oil}(eq) - C_{wall}(eq)$ (wt%)	0.45	0.36	0.27	0.17
$J_{wax} \times 10^{10}$ (m/s·wt%)	19.25	16.20	13.18	10.12

Source: Huang, Z. et al., *Energy & Fuels*, 25, 5180–5188, 2011.

TABLE 6.4

Comparison of the Parameters Involved in the Characteristic Mass Flux of Wax Deposition, J_{wax}, among the Deposition Experiments with Different $T_{coolant}$ while Q_{oil} and T_{oil} Remained Constant

Parameters		Value	
T_{oil} (°C)		20.2	
$T_{coolant}$ (°C)	5.0	10.0	15.0
Q_{oil} (m³/h)		5.0	
T_{wall} (°C)	8.1	12.1	16.1
$D_{wax,wall} \times 10^{10}$ (m²/s)	1.93	2.44	2.97
C_{oil}(eq) (wt%)	1.26	1.26	1.26
C_{wall}(eq) (wt%)	0.48	0.89	1.13
C_{oil}(eq) – C_{wall}(eq) (wt%)	0.78	0.37	0.13
$J_{wax} \times 10^{10}$ (m²/s·wt%)	31.64	16.54	7.92

Source: Huang, Z. et al., *Energy & Fuels*, 25, 5180–5188, 2011.

characteristic mass flux, J_{wax}, and its associated parameters for this change in $T_{coolant}$ are shown in Table 6.4. It can be seen that the increase in $T_{coolant}$ leads to the increase in T_{wall}, which increased the values of only two parameters: D_{wo} and C_{wall}(eq). Here, the concentration of wax in the bulk oil, C_{oil}(eq), did not change as the oil temperature was not varied, as in the previous case. The increase in $D_{wax,wall}$ tends to increase the J_{wax}. However, the increase in C_{wall}(eq) from 0.48% to 1.13% caused a significant decrease in the mass transfer driving force, $[C_{oil}$(eq) – C_{wall}(eq)], from 0.78% to 0.13%. This decrease in the mass transfer driving force overcame the increase in the diffusivity, leading to a decrease in the characteristic mass flux with increasing $T_{coolant}$, as shown in the last row of Table 6.4. This decrease in the characteristic mass flux explains the reduced experimental deposit thickness, as shown in Figure 6.8.

In summary, the change in the characteristic mass flux, J_{wax}, in these experiments with varying temperature conditions is shown to be consistent with the change in the deposit thickness. It was further observed that the mass transfer driving force, $[C_{oil}$(eq) – C_{wall}(eq)], is the most influential parameter in determining the characteristic mass flux, and it should be ultimately the indicator for the impact of temperature on wax deposition.

6.4.3.2 Analysis on Another Oil

In Section 6.4.3.1, the concept of the characteristic mass flux, J_{wax}, was used to investigate the impact of operating temperature on wax deposition for a North Sea waxy condensate. In this section, the same analysis on another oil will be applied to investigate the validity of such methodology.

The study of Bidmus and Mehrotra (2009) was chosen in which a model wax–oil system was used in a mini flow loop (1 in. inner diameter and 4 in. long) to study the impact of oil temperature (T_{oil}) on wax deposition. Interesting results were found in this study: When T_{oil} is above the WAT, the deposit thickness decreases with increasing T_{oil}. When T_{oil} is below the WAT, the deposit thickness increases with increasing T_{oil}. While no mass transfer explanation was given in the original study to help explain such observations, from our previous analysis with the North Sea condensate, an investigation using the mass transfer driving force, [$C_{oil}(eq) - C_{wall}(eq)$], as well as the characteristic mass flux for wax deposition, J_{wax}, might generate some important insights. Consequently, the same solvent, Norpar13 from Imperial Oil (Ontario, Canada), and the same wax sample, Parowax, from Conros Corp. (Ontario, Canada), were purchased (Huang, Lu et al., 2011). The measured wax precipitation curve is shown in Figure 6.10. Comparing Figure 6.10 with Figure 6.3, one notes that the precipitation curve for the model oil–wax system is almost linear during the range of experimental conditions, while the solubility curve for the North Sea condensate exhibits more precipitation at lower temperatures.

The deposition experiments in the study of Bidmus and Mehrotra (2009) are summarized in Figure 6.11, and the trend will now be analyzed with the approach using the characteristic mass flux, J_{wax}. Using the same method of analysis as for the North Sea condensate, the values of J_{wax} and its associated parameters for the model wax–oil experiments are summarized in Table 6.5.

It can be seen that the characteristic mass flux, J_{wax}, is again consistent with the amount of deposit observed in Figure 6.11 in that they both increase with increasing oil temperature until the oil temperature reaches the WAT. After reaching the WAT, both J_{wax} and the amount of deposit decrease with increasing temperature. The remarkable consistency in the relative magnitude of these two parameters further confirms the effectiveness of the use of the characteristic mass flux and the mass transfer driving force as predictors for the amount of wax deposit on the wall.

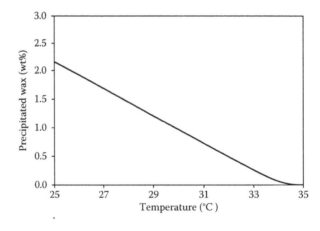

FIGURE 6.10
Solubility curve of wax for the model oil in the study of Huang, Lu et al. (2011). (Reprinted with permission from Huang, Z., Lu, Y., Hoffmann, R., Amundsen, L., & Fogler, H.S., *Energy & Fuels*, 25, 5180–5188. Copyright 2011 American Chemical Society.)

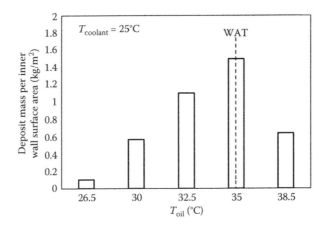

FIGURE 6.11
Comparison of the amount of deposit among experiments with different oil temperatures for the study of Bidmus and Mehrotra (2009). (From Bidmus, H.O., & Mehrotra, A.K., *Energy & Fuels*, 23, 3184–3194, 2009. Reprinted with permission from Huang, Z., Lu, Y., Hoffmann, R., Amundsen, L., & Fogler, H. S., *Energy & Fuels*, 25, 5180–5188. Copyright 2011 American Chemical Society.)

TABLE 6.5

Comparison of the Parameters for J_{wax} among Deposition Experiments with Different T_{oil} while Q_{oil} and $T_{coolant}$ Are Maintained Constant in the Study of Bidmus and Mehrotra (2009) Using a Model Oil–Wax System

Parameters			Value		
T_{oil} (°C)	26.5	29.0	33.0	35.0	38.5
$T_{coolant}$ (°C)			25.0		
Q_{oil} (m³/h)			0.4		
T_{wall} (°C)	25.4	25.9	27.1	28.0	29.6
$D_{wax,wall} \times 10^{10}$ (m²/s)	2.81	2.94	3.27	3.53	4.07
C_{oil}(eq) (wt%)	4.19	4.80	5.74	6.00	6.00
C_{wall}(eq) (wt%)	3.94	4.05	4.34	4.55	4.93
C_{oil}(eq) – C_{wall}(eq) (wt%)	0.25	0.75	1.40	1.45	1.07
$J_{wax} \times 10^{10}$ (m²/s·wt%)	27.66	86.81	180.24	201.52	171.28

Source: Huang, Z. et al., *Energy & Fuels*, 25, 5180–5188, 2011.

6.4.3.3 Importance of the Wax Precipitation Curve

The solubility curve has a great impact on the characteristic mass flux, J_{wax}, by affecting the concentration difference of wax, [C_{oil}(eq) – C_{wall}(eq)], at different temperatures. To visualize such impact, the changes in T_{wall}, C_{oil}(eq), and C_{wall}(eq) when T_{oil} was varied are shown in Table 6.6 for the two studies analyzed above, i.e., one with the North Sea condensate and the other with the model oil–wax mixture.

One observes that for both sets of the experiments, the increases in T_{oil} have caused the increases in T_{wall}. The increases in T_{wall} were less significant than

TABLE 6.6

Comparison of the Changes in T_{oil}, T_{wall}, $C_{oil}(eq)$, and $C_{wall}(eq)$ for the Experiments with the North Sea Oil A and the Model Oil–Wax Mixture with Varying T_{oil}

	North Sea Condensate	Model Oil–Wax Mixture
Change in T_{oil} (°C)	15.3–35.4	26.5–35.0
ΔT_{oil} (°C)	20.1	8.5
ΔT_{wall} (°C)	11.0	2.6
$\Delta T_{oil} > \Delta T_{wall}$	Yes	Yes
$\Delta C_{oil}(eq)$ (wt%)	0.35	1.81
$\Delta C_{wall}(eq)$ (wt%)	0.61	0.61
$\Delta C_{oil}(eq) > \Delta C_{wall}(eq)$	No	Yes

the increases in T_{oil} for both studies. The major difference is that for the model oil–wax mixture, the gradient of its solubility curve (Figure 6.10) is virtually constant (the solubility curve is close to a straight line), so that the changes in $C_{oil}(eq)$, and $C_{wall}(eq)$ simply reflected the changes in T_{oil} and T_{wall}. These changes eventually lead to the increase in the mass transfer driving force, $[C_{oil}(eq) - C_{wall}(eq)]$, and in the mass flux, J_{wax}, for the model mixture when T_{oil} increased. However, the gradient of the solubility curve for the North Sea condensate decreased with increasing temperature because the solubility curve is concave, as shown in Figure 6.3. This concave shape of the solubility curve for the North Sea condensate results in the change in $[C_{oil}(eq) - C_{wall}(eq)]$ to deviate from the change in $[T_{oil} - T_{wall}]$. This finding further demonstrates the advantage of using the mass transfer driving force in comparison to the thermal driving force as it includes the impact of the solubility curve on wax deposition.

6.4.3.4 Carbon Number Distribution of the Oil

The above comparison has highlighted the importance of the shape of the solubility curve on the effects of the oil/coolant temperatures on wax deposition. The solubility of wax represents the multicomponent solid–liquid equilibrium of the oil, which strongly depends on its carbon number distribution (CND). Figure 6.12 shows the CNDs of the n-C_{20+} components for the wax used in the model mixture and the distribution for the North Sea condensate. Two major differences can be observed between these two oils. First, a longer tail can be seen in the CND of the North Sea condensate, indicating the existence of heavy paraffins (from n-C_{50} to n-C_{80}), which are not existent in the model wax. In addition, the lighter components (from n-C_{20} to n-C_{26}) account for 41% of the wax in the North Sea condensate, while they only account for 26% of the model wax. The small amount of the heavy components and the excess amount of the light components in the North Sea condensate cause a greater precipitation of wax at lower temperatures compared to the model oil–wax mixture. This greater amount of precipitation is consistent with the

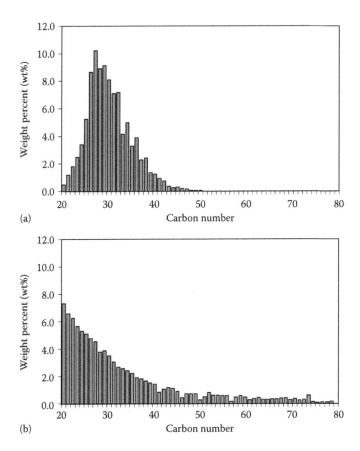

FIGURE 6.12
CND of the heavy components in (a) the model oil–wax mixture and (b) the North Sea condensate. (Reprinted with permission from Huang, Z., Lu, Y., Hoffmann, R., Amundsen, L., & Fogler, H. S. *Energy & Fuels, 25,* 5180–5188. Copyright 2011 American Chemical Society.)

concave shape of the solubility curve of the North Sea condensate at low temperatures and eventually explains the contradictory observed trends of growth of deposit thickness between Figures 6.7 and 6.11.

6.5 Applying Wax Deposition Models to Investigate Deposit Compositions

In Section 6.4, the behavior of the deposit thickness at different operating temperatures was analyzed using the first principles of transport phenomena. However, equally important is the composition in the wax deposit, which not only makes a quantitative evaluation on the degree of diffusion of the waxy components possible but also provides critical information for the strength of the wax deposit for the design of techniques for deposit removal (Bai & Zhang, 2013a). The composition of the waxy deposit can be characterized by its CND. Unfortunately, it is virtually impossible to determine the CND of a wax deposit formed at a particular section of a subsea pipeline. The deposit's CND can be predicted by coupling wax thermodynamic modeling and wax deposition modeling to obtain a fundamental understanding of the behavior of waxy components with different CNDs.

As discussed in detail in Chapter 4, Sections 4.1 and 4.2, it is widely accepted that molecular diffusion is the main mechanism responsible for deposit formation. Each alkane component diffuses to the wall due to its radial concentration gradient. Depending on the carbon number, different n-alkanes will have different physical properties, e.g., solubility values and molecular diffusivities, and thus can have different concentration gradients. This variation in the concentration gradient for different n-alkanes could result in a difference in the composition of the n-alkanes in the deposit formed by the differences in molecular diffusion of multiple n-alkane components. According to the study by Zheng, Zhang, Huang, and Fogler (2013), the solubility differences and the resulting concentration driving force difference for molecular diffusion dominate the evolution of the CND. The concentration driving force for precipitating paraffin component can be calculated by taking the difference of their bulk and wall concentrations: $\Delta C_i = C_{i,\text{bulk}}(T_{\text{bulk}}) - C_{i,\text{wall}}(T_{\text{wall}})$. The bulk and wall concentrations of a paraffin component, i.e., $C_{i,\text{bulk}}(T_{\text{bulk}})$ and $C_{i,\text{wall}}(T_{\text{wall}})$, will be evaluated at their corresponding equilibrium concentrations in the liquid phase at the bulk and wall temperatures, i.e., T_{bulk} and T_{wall}. These two concentrations can be calculated using wax thermodynamic models. By using wax thermodynamic modeling techniques, Zheng et al. (2013) calculated the concentration driving force distribution: ΔC_i as a function of carbon number i. As the ΔC_i distribution dominates the evolution of deposit CND, the deposit formed in two different experiments, such as those listed in Table 6.7, can have a similar CND as long as the concentration driving force distributions are similar.

Figure 6.13 shows the comparison between the calculated concentration driving force distribution and the observed deposit CND in these two experiments with significantly different operating conditions.

TABLE 6.7

Summary of Two Sets of Operating Conditions:
A and B Resulting in Similar Deposit CND

	Condition A	Condition B
Q_{oil} (m³/h)	5.00	21.00
T_{oil} (°C)	20.17	15.24
T_{coolant} (°C)	5.00	10.00
T_{wall} (°C)	8.39	12.49

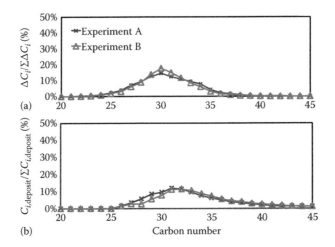

FIGURE 6.13

(a) Comparison of the concentration driving force distribution between conditions A and B and (b) comparison of the deposit CND generated by two sets of operating conditions. (Reprinted with permission from Zheng, S., Zhang, F., Huang, Z., & Fogler H. S., *Energy & Fuels, 27*, 7379–7388. Copyright 2013 American Chemical Society.)

As can be seen from Figure 6.13a, theoretical analysis by thermodynamic modeling indicates a similar concentration driving force distribution under the two sets of experimental conditions. The experimentally observed deposit CNDs in the two experiments being similar to each other, as shown in Figure 6.13b, verified the theoretical analysis. Consequently, in order to predict the deposit CND under a particular set of operating conditions, one only needs to apply the following three-step procedure:

- Use transport modeling to determine the wall and centerline temperatures
- Use thermodynamic modeling to determine the equilibrium concentrations of waxy components at the wall and centerline temperatures
- Calculate the concentration driving force distribution based on the equilibrium concentrations of the waxy components

The deposit CND under the particular set of the operating conditions should resemble the concentration driving force distribution.

6.6 Summary

In this chapter, the performance of a model developed at the University of Michigan, known as the Michigan Wax Predictor (MWP), is highlighted based on a series of wax deposition experiments with a wide variety of operating conditions. The evaluation was first carried out based on the comparison of the wax deposit thickness between experimental measurements and model predictions. It was found that the MWP was able to predict the following trends *without* applying any tuning parameters:

- Decrease in the deposit thickness with increasing oil flow rate
- Decrease in the deposit thickness with increasing oil temperature
- Decrease in the deposit thickness with increasing coolant temperature

Based on this consistency, the transport equations in the MWP are carefully examined to identify the reasons for such agreements. With the approach of a dedimensionalized analysis, a representative parameter to highlight the degree of wax deposition is proposed: the characteristic mass flux for wax deposition:

$$J_{wax}(t = 0) = \frac{D_{wax,wall}\left[C_{oil}(eq) - C_{wall}(eq)\right]}{R_{pipe}} \tag{6.10}$$

Within this parameter, the effect of temperature can be separated into two categories: The first is on the transport properties of the oil (to affect the wax diffusion coefficient at the wall temperature, $D_{wax,wall}$), and the second is on the thermodynamic properties of the oil (to affect the wax equilibrium concentrations at the bulk oil and at the wall). It was found that in most of the wax deposition experiments, the effect of temperature on the thermodynamic properties often outweighs the effect on the transport properties, so that the change in the deposit thickness is often found to be consistent with the change in the concentration difference, $[C_{oil}(eq) - C_{wall}(eq)]$. Such difference is considered to be the mass transfer driving force for deposition, and it is an important indicator on how the deposition behavior varies when the operating conditions are altered. This methodology was further validated with a wax–oil model mixture in a different deposition flow loop. The consistency of this methodology for different oils and different experimental apparatuses has strategic importance for model scale-up for field predictions. In fact, in Chapter 7, it will be seen that the MWP is found to be able to achieve satisfactory agreement with field data.

In addition to the comparison of the deposit thickness, an analysis is made based on the comparison of the composition of the wax deposit between experimental characterization and model predictions. In this case, wax deposition modeling by the MWP is combined with a thermodynamic model (Coutinho & Stenby, 1996; Coutinho & Ruffier-Méray, 1997; Coutinho, 1998; Coutinho & Daridon, 2001; Coutinho, Edmonds, Moorwood, Szczepanski, & Zhang, 2006) to resolve the mass transfer of alkanes with different chain length. It was discovered that the evolution of deposit composition is also dominated by the distribution of concentration driving forces $[C_{oil}(eq) - C_{wall}(eq)]$ among alkanes with different chain length.

7

Applying Wax Deposition
Models for Field Predictions

7.1 Introduction

7.1.1 Wax Control Strategies for the Field

In this chapter, thermodynamic modeling and transport modeling are used to help develop remediate strategy characteristics for wax deposition in the field. A general methodology of wax deposition evaluation and control is highlighted in Figure 7.1.

The first step to assess the possibility of wax deposition is to determine the wax appearance temperature (WAT). The WAT is usually compared with the minimum wall temperature of the pipeline during operations. The wall temperature of the pipeline can be calculated by most flow simulators that have been used extensively for pipeline designs for the past decades. Depending on the values of the WAT and the minimum wall temperature, the following possible scenarios can occur:

- *Scenario 1*: Wax deposition would probably not pose a risk for normal operations if the WAT is significantly lower than the minimum wall temperature. However, one should be on the lookout for situations where the flow rate of the fluids decreases (production turndown), causing the pipeline wall temperature to fall below the WAT.

- *Scenario 2*: If the minimum wall temperature of the pipeline is close to or only slightly below the WAT, there can be a risk of wax deposition in the pipeline. In this case, the problem of wax deposition might be addressed by increasing the insulation or installing heating facilities (for example, direct electrical heating or pipe-in-pipe heating) over the section of the pipe where the wall temperature might drop to below the WAT. For short pipelines, insulation and heating methods might still be economically viable even if the minimum wall temperature is significantly below the WAT.

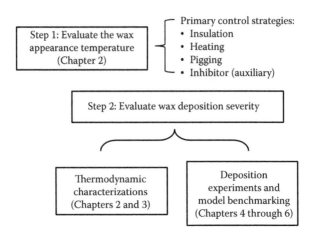

FIGURE 7.1
Summary of general wax deposition screening and control methodologies.

- *Scenario 3*: The worst case is in long subsea pipelines where the minimum wall temperature falls significantly below the WAT. In this case, pipeline insulation and heating either are too costly or cannot be installed, and wax control strategies go from preventive methods to mitigation methods. The most common mitigation method is pigging, where an inspection gauge is sent through the pipeline periodically to mechanically remove the wax deposits. The word "pigging" might originate from "*p*ipeline *i*ntervention *g*adget" or refer to the sound generated during the travelling of the device. Because normal operations often need to be paused during the operation of pigging, the frequency of pigging greatly affects the cost of production (as previously shown in Figure 1.2). In order to determine an economical pigging frequency, deposition modeling is often needed as the next step (step 2 in Figure 7.1) to understand the severity of wax deposition and to establish an appropriate pigging frequency.

It should be noted that wax concerns in the subject of flow assurance are not limited to deposition during production. During pipeline shutdowns, the flow is stopped, and the fluid is being cooled to the ambient temperature. In this case, if the cool down time is too long, the waxy components can also precipitate from the oil and cause gelation to occur in the pipeline. Wax gelation normally occurs during static conditions (pipeline shutdown) in which the entire fluid (from the wall to the pipe centerline) can be gelled, while wax deposition mainly occurs on the pipe wall during flow conditions. Although

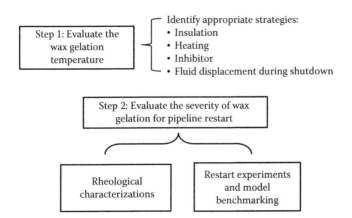

FIGURE 7.2
Summary of general wax gelation screening and control methodologies.

not discussed in detail in this book, the control strategies of wax gelation are analogous to those for wax deposition, as summarized in Figure 7.2.

7.1.2 Evaluating the Severity of Wax Deposition: The Ideal vs. the Reality

For the step to investigate the severity of wax deposition (step 2 in Figure 7.1), a variety of experimental measurements and modeling techniques have been discussed in Chapters 2 through 6. For example, these characterization methods include determining the n-paraffin distributions using high-temperature gas chromatography (HTGC), measuring the WAT by cross-polarized microscopy (CPM), determining the wax precipitation curve using differential scanning calorimetry (DSC), etc. During the process of thermodynamic modeling, the properties of the waxy components often need to be adjusted until the model predictions match the experimental results. In addition, flow-loop wax deposition experiments should be carried out to benchmark a deposition model prior to its application for field predictions.

However, due to the cost, time, and other concerns for inconvenience, industrial users often focus on a few measurements that come from years of flow assurance management experience. For example, one might find that the method of CPM delivers the most accurate measurement of the WAT, whereas another might believe that other methods, such as DSC, provide an estimation that is enough for field design. In Sections 7.2 and 7.3, two published examples of field applications will be discussed in which different laboratory characterizations are used to form the basis for field predictions.

7.2 Example 1—Single-Phase Pipe Flow

7.2.1 Introduction

The first field study example of wax deposition is based on the study by Singh, Lee, Singh, and Sarica (2011). In this study, a crude oil from offshore Indonesia was transported between a central processing platform and floating production storage and offloading (FPSO) in a 23-km-long single-phase subsea pipeline. Wax deposition was found to be the primary flow assurance concern, and pigging was applied as the main mitigation method. The crude oil, with an American Petroleum Institute (API) gravity of 45, has a WAT of 58°C measured from CPM (Figure 7.3).

The operating conditions and the oil properties are summarized in Table 7.1. The inlet temperature is around 75°C. With a transport pipeline of 23 km, it is expected that the wall temperature could eventually decrease to below

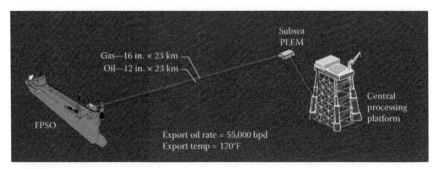

FIGURE 7.3
Field layout in the study of Singh et al. (2011). (From Singh, A. et al., Flow assurance: Validation of wax deposition models using field data from a subsea pipeline. In *Offshore Technology Conference* [pp. 1–19]. Houston, TX: Offshore Technology Conference, 2011.)

TABLE 7.1

Summary of Operating Conditions in the Study of Singh et al. (2011)

Pipeline diameter	12 in.
Pipeline length	23 km
External heat transfer coefficient (insulation included)	22 W/m²/K
Oil flow rate	300,000 bpd
Inlet temperature	75°C
Outlet pressure	350 psig
Predeposition pressure drop	200 psi
Pigging frequency	Weekly

Source: Singh A. et al., Flow assurance: Validation of wax deposition models using field data from a subsea pipeline. In *Offshore Technology Conference* (pp. 1–19). Houston, TX: Offshore Technology Conference, 2011.

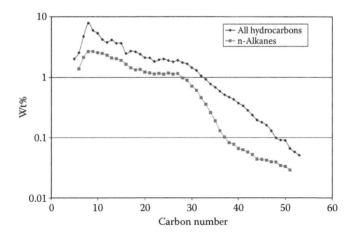

FIGURE 7.4

Composition of the oil and its n-paraffin distribution. (From Singh, A. et al., Flow assurance: Validation of wax deposition models using field data from a subsea pipeline. In *Offshore Technology Conference* [pp. 1–19]. Houston, TX: Offshore Technology Conference, 2011.)

the WAT of 58°C. The composition and the n-paraffin distribution of the oil are shown in Figure 7.4. Based on this analysis, the oil has 17% n-C_{19+}, which is generally known as the waxy components. Other detailed properties of the oil can be found in the original study of Singh et al. (2011).

7.2.2 Wax Thermodynamic Characterizations

In the original study, thermodynamic modeling was carried out utilizing the measured n-paraffin distribution using the model of Erickson, Niesen, and Brown (1993), which is implemented in the program of the wax deposition model from the University of Tulsa (TUWAX). Here, another thermodynamic model was used—Coutinho's Wilson model (Coutinho & Ruffier-Méray, 1997). This model is implemented in the program of MultiFlash 4.4. The improvement of Coutinho's model over the Erickson model is that it accounts for the nonideality of the solid phase formed by the waxy components. A detailed comparison of these two models was discussed in Chapter 3, Sections 3.2 and 3.3.

Tuning is often made to match the predicted precipitation curve with one or a few experimental data points. The process of tuning the thermodynamic models is discussed in detail in Chapter 3, Section 3.5. Here, the predicted wax precipitation curve using Coutinho's model was tuned to the point of WAT of 58°C with an onset precipitation fraction of 0.045 wt%. This onset fraction refers to the amount of wax precipitation that can be detected by the measuring device. The value of 0.045 wt% is the value recommended by the software vendor based on a general detection limit of CPM as the WAT measuring method. During tuning, the program of MultiFlash adjusted the

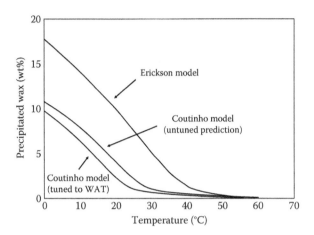

FIGURE 7.5
Comparison of the wax precipitation curves from different thermodynamic models.

values of the heat of crystallization of the waxy components to match the point of 0.045 wt% wax precipitation at 58°C.

A comparison of these wax precipitation curves is shown in Figure 7.5. The reported n-C_{19+} fraction from the n-paraffin distribution is 17%. It is not likely that all of the n-C_{19+} have precipitated at 0°C because some of the lighter components (for example, n-C_{19} and n-C_{20}) are expected to continue to precipitate at even lower temperatures. Therefore, the prediction given by the Erickson model that more than 17% of waxy components have precipitated at 0°C is likely to be an overestimate. This overestimate in the amount of wax precipitation might be associated with the fact that the Erickson model considers the ideal solutions for the wax solid phase, whereas the Coutinho model accounts for the nonideality in the solid solutions. The details of the thermodynamic assumptions for these two models are highlighted in Tables 3.1 and 3.2. Consequently, the prediction based on the Coutinho model that is tuned with the WAT is used as the wax precipitation curve for this study.

7.2.3 Deposition Predictions and Pigging Frequency Design

In this section, a wax deposition model called the Michigan Wax Predictor (MWP) is used to develop field predictions for this study (Huang, Lee, Senra, & Fogler, 2011; Lee, 2008). As discussed in Chapter 6, the MWP has shown outstanding performance in modeling laboratory deposition experiments by correctly predicting the change in deposition rates as a function of operating conditions without the use of any tuning parameters. In this chapter, the same model configuration (no adjustable parameter) in the MWP for the field predictions will be applied.

Figure 7.6 shows the "predeposition" bulk and the inner wall temperature profiles calculated by the MWP, which are the temperature profiles before

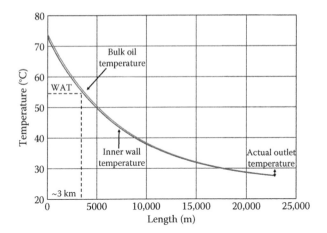

FIGURE 7.6
Prediction of the temperature profiles in the pipeline.

any wax has deposited. In fact, it is very important to compare the predicted bulk temperature of the fluid with field measurements if possible to confirm the validity of the heat transfer calculation by the wax model. It is seen that due to the insulation (external heat transfer coefficient of 22 W/m/K) of the pipeline, the bulk and the wall temperature profiles are quite similar. The temperature profiles are consistent with field measurements, which have recorded an outlet temperature of 27°C–29.5°C. The agreement between the calculated temperature and the measured value is critical to develop reliable deposition predictions. The location where the wall temperature drops to below the WAT is around 3 km downstream of the inlet, which indicates that even with insulation, approximately 20 km of the pipeline is subject to the risk of wax deposition.

After gaining insights from the temperature profile along the pipe, we are ready to investigate the growth of the wax deposit thickness. The visualization of deposit thickness from field predictions is slightly different from that during the modeling of laboratory deposition experiments: In previous discussions in Chapter 6 about the laboratory experiments, the evolution of axial averaged deposition thickness (*x*-axis) as a function of time (*y*-axis) was shown. Because for the laboratory experiments, the length of the pipe is much shorter, the deposit profile throughout the pipe is expected to be uniform. This axial averaged deposit thickness was compared with the experimental measurements using the pressure drop method, i.e., the pressure-drop of the pipe was used to determine the effective radius of the pipe. The difference of the effective radius and the pipe inner radius is the deposit thickness. This method of plotting was also used in many other laboratory wax deposition studies (Hernandez, 2002; Hoffmann & Amundsen, 2010; Huang, Lu, Hoffmann, Amundsen, & Fogler, 2011; Lund, 1998; Singh et al., 2000; Venkatesan, 2004).

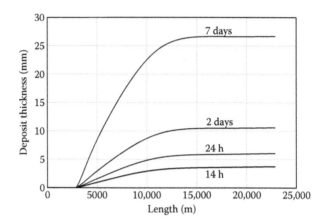

FIGURE 7.7
Predictions of the deposit thickness along the pipeline at different times by the MWP.

However, this method of visualization might not be suitable when it is applied to the field. In the current analysis on wax deposition on field pipelines, the variation of deposit thickness along the pipe can be significant. Therefore, the averaged deposit thickness over the entire pipeline has much less physical significance. In these laboratory studies, the evolution of axial averaged thickness (plotted in the y-axis) versus time (plotted in the x-axis) is often shown for analysis. However, for predictions based on field pipelines, the variation of deposit thickness along the pipe can be much more significant. Therefore, the axial averaged deposit thickness has much less physical significance for field predictions than for laboratory modeling. Consequently, instead of plotting the axial averaged deposit thickness as a function of time, two other methods to visualize the deposition predictions and compare with field measurements were used: First, the axial profile of the deposit thickness is plotted, as shown in Figure 7.7. It is seen that after 7 days, the maximum deposit thickness of the pipe reached 27 mm (equivalent to one tenth of the pipeline inner diameter).

To further investigate the credibility of this prediction, the pressure drop of the entire pipeline during the course of wax deposition is calculated based on the Blasius friction factor correlation (Wilkes, 2005), as shown in Equation 7.1. To calculate the pressure drop of the entire pipeline, the pipeline is discretized into small segments, and the pressure drop of each segment was first calculated based on the effective diameter of the section, as shown in Equation 7.2. Then, the pressure drop of the entire pipeline can be determined by adding up the pressure drop of all the segments of the pipe, as shown in Equation 7.3.

$$f_{Darcy} = 0.316 \, Re^{-0.25} \tag{7.1}$$

$$\Delta P_{pipe, \, i} = f_{Darcy} \frac{L_{pipe, \, i}}{d_{effective}} \frac{\rho_{oil} U^2}{2} \tag{7.2}$$

$$\Delta P_{\text{pipe}} = \sum_i \Delta P_{\text{pipe},\,i} \tag{7.3}$$

This calculation was performed during the course of wax deposition so that, eventually, the increase in pressure drop predicted by the MWP could be obtained, as highlighted in Figure 7.8 in comparison with field measurements. It is seen that without any tuning and laboratory benchmarking, the MWP provides reasonable prediction for the pressure-drop increase due to wax deposition with slight overprediction at later times.

In this study, pigging is used as the primary remediation method for wax deposition. For the design of pigging frequency, the time for the maximum thickness of the deposit in the pipe to reach a certain threshold is often used. This threshold is designed to prevent the pig from getting stuck, and it depends on many factors, including the type of the pig, the applied pressure for pigging, the properties of the fluid used to propel the pig, the efficiency of the previous pigging and the contingency plan in case the pig gets stuck in the pipe. Typically, the establishment of an appropriate value for this threshold requires significant operational experience. Generally, the industrial preference of such threshold is around 4 mm for field design, which in many cases can be conservative (Golczynski & Kempton, 2006). During field production, operators in the field might decrease the frequency of pigging if they believe that the original suggested frequency is overconservative.

Here, we do not have the intention to establish a general pigging frequency threshold because of the limited data available. However, the pigging frequency that was actually carried out in this field can be used to see the value of the corresponding threshold for maximum deposit thickness in this case.

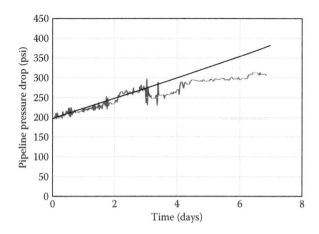

FIGURE 7.8
Comparison of pipeline pressure drop during the course of wax deposition between MWP predictions and field measurements.

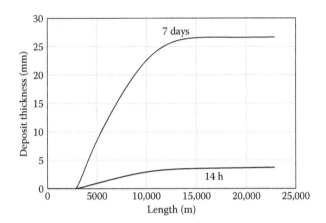

FIGURE 7.9
Prediction of deposit thickness along the pipe after 7 days.

For this study, it was eventually found that pigging could be performed every 7 days without causing any operational concerns (Singh et al., 2011). After 7 days, the axial distribution of the deposit thickness predicted by the MWP is 27 mm, as previously shown in Figure 7.7. In other words, in this study, if the pigging frequency is determined by using the deposition model of MWP, the threshold for the maximum thickness allowed could be as much as 27 mm. In this case, applying the general value of 4 mm seems conservative for this case, and the pigging frequency would change to every 14 h, which in hindsight seems conservative (Figure 7.9).

7.3 Example 2—Gas/Oil Flow

7.3.1 Introduction

The second application example focused on the development in Well B, Campos Basin in Brazil operated by Petrobras at a water depth of approximately 800 m (Noville & Naveira, 2012). The multiphase petroleum mixture (oil and gas) was transported from the wellhead to the topside via a 5-km flow line and a riser. A schematic of the production pipeline is shown in Figure 7.10.

The minimum wall temperature of the flow line was expected to fall below the WAT of the oil, which is approximately 17°C. The oil has an API gravity of 26.6, and a downhole gas lift was applied to facilitate the production. The operating conditions are summarized in Table 7.2. In this study, wax modeling was carried out using two commercialized programs: PVTsim, an industrial thermodynamic modeling package, was used for wax thermodynamic modeling. OLGA, an industrial multiphase flow simulator, was used

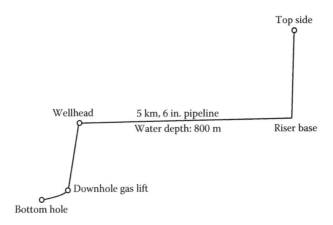

FIGURE 7.10
Schematic of the production pipeline for the study of Noville and Naveira (2012).

TABLE 7.2

Summary of the Operating Conditions in the Study of
Noville and Naveira (2012)

Pipeline diameter	6 in.
Pipeline length	5 km
Oil flow rate	2617 bpd
Gas oil ratio (GOR)	618
Lift gas flow rate	5.8 mmscfd
Reservoir temperature	78°C
Reservoir pressure	2217 psig
Topside pressure	155 psig
Pigging frequency	14 days

Source: Noville, I. & Naveira, L., Comparison between real
field data and the results of wax deposition simulation. In *SPE Latin America and Caribbean Petroleum
Engineering Conference* (pp. 1–12). Mexico City, Mexico:
Society of Petroleum Engineers, 2012.

for wax deposition modeling. The detailed descriptions for the wax thermodynamic model in PVTsim can be found in Chapter 3, Section 3.3. The
detailed transport equations for the wax deposition model in OLGA can be
found in Chapter 4, Sections 4.2 and 4.3.

7.3.2 Wax Thermodynamic Characterization

Different from Example 1 discussed in Section 7.2, the n-paraffin distribution
in this study is not measured in this example in order to carry out thermodynamic modeling to obtain the wax precipitation curve. Instead, the wax
precipitation curve was directly measured by the authors via the DSC.

The next step is to import this information of the wax precipitation curve to a wax deposition model. However, here emerges a problem about industrial practicality: The wax deposition model in the industrial flow simulator OLGA only accepts the input of a wax precipitation curve in a format that is developed by selected industrial thermodynamic models that it supports. In this study, the program of PVTsim is used for this purpose. The program of PVTsim generates a "wax table" that consists of the amount of wax precipitated at different temperatures and pressures. This wax table could be directly used by OLGA for wax deposition simulations. In this case, the n-paraffin distribution is adjusted in a thermodynamic model order to tune the predicted wax precipitation curve to match the DSC results. The thermodynamic model used is based on the study by Pedersen, Thomassen, and Fredenslund (1985). However, we can see that this information is not as important here as the thermodynamic model was only used as a dummy. The real source of the precipitation curve is the DSC experiment. The DSC measurements as well as the thermodynamic model fitting are shown in Figure 7.11. It is seen that the predicted precipitation curve matches the experimental measurements from 0°C to 17°C. However, at the range of 17°C–35°C, there seems to be a very small amount of wax precipitation that was underpredicted by the thermodynamic model.

This method of thermodynamic characterization indicates that the method of DSC was preferred in this study to study the wax precipitation curve, which is different from the previous example (Example 1) where thermodynamic prediction was carried out based on the n-paraffin distribution measured by the HTGC. A comparison of these two strategies is highlighted in Table 7.3.

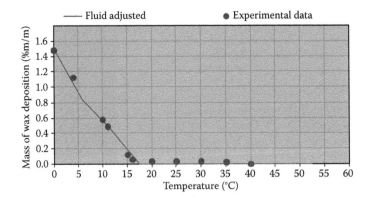

FIGURE 7.11
Experimental measurement of the wax precipitation curve and the thermodynamic fitting in the study of Noville and Naveira (2012). (From Noville, I. & Naveira, L., Comparison between real field data and the results of wax deposition simulation. In *SPE Latin America and Caribbean Petroleum Engineering Conference* [pp. 1–12]. Mexico City, Mexico: Society of Petroleum Engineers, 2012.)

TABLE 7.3

Comparison of the Two Thermodynamic Characterization Strategies Used in Examples 1 and 2

	Example 1	Example 2
Experiments carried out	n-Paraffin distribution using HTGC WAT using CPM	Entire wax precipitation curve using DSC
Purpose of using thermodynamic modeling	To predict a wax precipitation curve to be used by a deposition model	To generate a wax precipitation curve same as that measured by DSC for a deposition model
Parameters to adjust during tuning the thermodynamic model	A different thermodynamic model offers different tuning options, for example, the heat of crystallization and the detection limit of the WAT measuring device. A complete list for various models was discussed in Table 3.2	The n-paraffin distribution
Parameters to match during modeling	WAT measured using CPM	Entire wax precipitation curve measured by DSC

7.3.3 Deposition Predictions and Pigging Frequency Design

The presence of oil and gas multiphase flow greatly increases the level of complexity in terms of fluid mechanics and heat/mass transfer. In this case, the deposition modeling with the MWP is not included, as the composition of the production fluid as well as the detail pipeline geometry has not been disclosed in the original study. Without these pieces of information, it would be difficult to determine the hydrodynamics and heat transfer characteristics in the pipeline accurately. Therefore, we chose to simply highlight the deposition modeling results published in the original study by Noville and Naveira (2012) using OLGA wax and discuss the associated uncertainties.

During the course of wax deposition, the increase in the pressure drop across the pipeline is measured via the downhole pressure gauge, while the pressure at the topside outlet remains constant (155 psig). A comparison of the increase in the downhole pressure predicted from the OLGA wax simulator and the field measurements is shown in Figure 7.12. Several observations can be made: First, the measured downhole pressure does not always start at the lowest value (~1260 psig), which indicates that there is a variation in the degree of removal of the wax deposit during each pigging (this observation is also reported in the original study).

Second, the original model prediction with a default value of the wax diffusion coefficient significantly underestimates the increase in the downhole pressure of the pipeline. In this case, the diffusion coefficient, D_{wax}, was multiplied by a factor of 100 to represent the highest rate of increase (the slope of the curves) in the downhole pressure and by a factor of 22 to represent the lowest increase rate as the pessimistic and optimistic cases, respectively. The wax diffusion coefficient estimated from the Hayduk–Minhas correlation for n-paraffin solutions was based on as many as 58 data points (Hayduk &

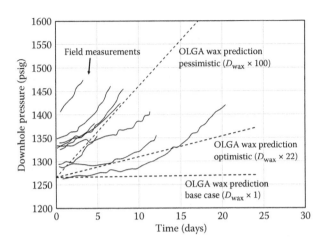

FIGURE 7.12
Comparison of the increase in the downhole pressure between OLGA wax predictions and field measurements. (From Noville, I. & Naveira, L., Comparison between real field data and the results of wax deposition simulation. In *SPE Latin America and Caribbean Petroleum Engineering Conference* [pp. 1–12]. Mexico City, Mexico: Society of Petroleum Engineers, 2012.)

Minhas, 1982). Consequently, significant tuning might lead to unrealistic wax diffusion coefficients, and the mismatch between the base case (no multiplier on D_{wax}) and the field measurements might be due to uncertainties in other factors, such as the wax precipitation curve. Consequently, more laboratory characterizations and benchmarking (as introduced in Section 7.1.1) might be beneficial to identify the discrepancies.

The prediction of the deposit thickness after 14 days is shown in Figure 7.13 (Well B). It can be seen that the maximum deposit thickness reaches

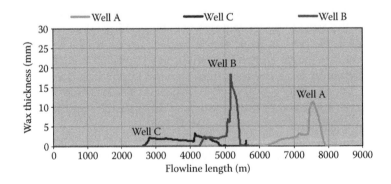

FIGURE 7.13
Predictions of deposit thickness using the OLGA wax based on the optimistic scenario (D_{wax} multiplied by 22). (From Noville, I. & Naveira, L., Comparison between real field data and the results of wax deposition simulation. In *SPE Latin America and Caribbean Petroleum Engineering Conference* [pp. 1–12]. Mexico City, Mexico: Society of Petroleum Engineers, 2012.)

around 18 mm in 14 days. The pigging frequency is determined based on the following two criteria:

- The time for the predicted maximum deposit thickness along the pipe to reach a certain threshold
- The time for the predicted maximum mass of the wax deposit in the entire pipeline to reach a certain threshold

These two thresholds are designed to ensure that the pig does not become trapped in the pipeline during pigging operations. The final pigging frequency is based on the maximum number of days that satisfies both of these criteria. According to the original study, the threshold for the maximum predicted deposit thickness is reported to be 11 mm, and that for the maximum mass of wax deposit acceptable is 750 kg. It was found that the time for the first criterion is 7 days, and the time for the second criterion is 12 days. Consequently, the proposed pigging frequency was 7–10 days. The field was operating with this pigging frequency initially, and the pigging frequency eventually increased to once every 14 days.

7.4 Summary

7.4.1 Wax Thermodynamic Characterizations

One of the differences in the two examples in this chapter is the method used to characterize the wax precipitation curve. The first study by Singh et al. (2011) is based on the n-paraffin distribution measured by the HTGC. The wax precipitation curve is obtained from the prediction from a thermodynamic model, which is tuned to match the WAT, which was measured experimentally. In the second study by Noville and Naveira (2012), the wax precipitation curve is directly measured, and the use of a thermodynamic model is simply to generate the output of the precipitation curve for the input of a deposition model. In this case, the n-paraffin distribution is adjusted to fulfill the purpose. These two different methodologies reflect different levels of confidence on the wax thermodynamic characterization techniques, as summarized and shown in Table 7.4.

It should be noted that uncertainties are associated with the technologies used in these two approaches. For example, the n-paraffin distribution from HTGC is obtained by integration of the area under the peaks in a chromatogram, which can depend on the integration method (either the baseline method or the valley-to-valley method). The WAT detected by the CPM can be sensitive to its detection limit and the DSC measurements have been reported to have difficulties to differentiate between the noise and the signal for low wax content oils (such as the one for Example 2). A detail of the

TABLE 7.4

Comparison of the Two Wax Thermodynamic Characterization Methods Used in the Two Different Examples in This Chapter

Examples	n-Paraffin Distribution	Parameter to Match during Tuning	Measuring Technologies That the Method Relies On
1	Measured experimentally	WAT	HTGC, CPM Wax thermodynamic models
2	Adjustable parameters	Wax precipitation curve	DSC

limitations for various techniques for wax thermodynamic characterization can be seen in Chapters 2 and 3.

7.4.2 Wax Deposition Modeling

The above field application examples exhibit two different levels of complexities for wax deposition modeling. In the first example, the petroleum mixture is separated at a central processing facility prior to the flow toward the FPSO (Singh et al., 2011). In this case, the flow is single-phase oil flow, and wax deposition modeling becomes relatively simple. With the well-established single-phase hydrodynamics and heat/mass transfer correlations, the MWP was able to have reasonable agreement in predicting the increase in pressure drop during the course of wax deposition without any tuning. Comparisons of the pigging frequency recommended from model predictions with the actual pigging frequency in the field show that a general threshold of 4 mm maximum deposit thickness allowed can be overconservative.

The second example features a much more complicated scenario where the oil and gas multiphase mixture was transported together from the well to the topside processing facilities (Noville & Naveira, 2012). In this study, with the absence of extensive laboratory thermodynamic characterization and benchmarking, the uncertainties for wax deposition modeling greatly increase. The deposition model OLGA wax was not able to predict the increase in pressure drop across the pipe without tuning the wax diffusion coefficient significantly. The threshold for maximum deposit thickness is 11 mm for this case. However, with a different deposition model and model configurations, the appropriate threshold is likely to change.

7.5 Future Outlook

Based on the above field applications, it can be seen that despite several decades of wax deposition research, today, the application of wax modeling for field design still faces significant uncertainties. Consequently, industrial preference for the threshold of maximum deposit thickness is often over-conservative to account for these uncertainties. The standard practice currently is to use the industrial preference (on the order of 4 mm based on the study by Golczynski and Kempton [2006]) for the design stage of a field. During actual operations, the pigging frequency can be increased based on the inspection of the actual amount of wax deposits after each pigging. Still, overconservative pigging recommendations can lead to overconservative investment decisions that are often based on the analysis on the economic feasibility of the field during the design stage. Improving the rigorousness of wax deposition models is of critical importance to reduce the potential overconservatism in wax control strategies and to improve the understanding on the economic potential of a field. As discussed subsequently, there are several approaches that can potentially help to achieve this goal.

7.5.1 Improving Field Characterization Techniques

One of the most important and direct methods to evaluate a wax deposition model is to directly compare with field data. However, compared to laboratory experiments, which are often well designed and controlled, parameters in the field are much more difficult to monitor. Therefore, the advancements of the field measurement techniques can be of great potential to enable the measurement of many parameters that are currently not possible to assess in the field. For example, direct measurements of the wax deposit thickness of the pipe that do not interrupt normal productions would be very helpful to evaluate the severity of wax deposition in real time and to benchmark wax deposition models.

7.5.2 Collaboration between Industrial Partners

The development of appropriate wax modeling and control strategies is often based on operational experience accumulated over a long period of time and based on multiple petroleum field developments. Consequently, it could be beneficial for industrial members to share their wax modeling and control experience in order to have a more comprehensive understanding about the uncertainties in these strategies and potentially reduce the level of overconservatism in wax deposition management. Joint industrial technology research consortiums such as DeepStar and Research Partnership to Secure Energy for America (RPSEA) have carried out several industrial research projects on wax precipitation and deposition (DeepStar, 2011; RPSEA, 2007).

7.5.3 Develop More Rigorous Wax Deposition Modeling for Multiphase Flow Conditions

While for field study Example 1 (single-phase oil flow), the wax model without applying any tuning is able to achieve reasonable agreement on the increase in the pressure drop of the pipeline during the course of wax deposition, for Example 2, the agreement could not be reached without drastic tuning. The main reason for this discrepancy is on the occurrence of multiphase oil–gas flow. In addition to oil–gas flow, oil–water two-phase flows and oil–gas–water three-phase flows can also occur in field pipelines. Very few experimental studies attempt to understand the effects of the interplay between phases on wax deposition from fundamental perspectives of view. Moreover, the oil–gas–water three-phase wax deposition study has not yet been conducted. Lack of experimental characterization poses difficulties on the development of fundamental wax deposition models for multiphase flows. A variety of research directions are being undertaken at the University of Michigan in order to understand and model wax deposition in multiphase flows from first principles. As an example, in the case of water–oil flows, emulsified water can greatly change the rheological behavior of waxy crude oil, while it is still unclear how the rheology of waxy oil emulsion should be considered in any wax deposition models. The existence of a second phase in addition to the oil phase may also change the interaction between the flow and the pipe wall. In the case of water–oil two-phase flows, the wax deposition characteristics might also depend on the hydrophilicity and hydrophobicity of the pipe wall surface. Research and developments to understand the multiple perspectives of wax deposition characteristics in multiphase flows are essential to improving the efficiency in the numerical modeling.

Bibliography

Agrawal, K. M., Khan, H. U., Surianarayanan, M., & Joshi, G. C. (1990). Wax deposition of Bombay high crude oil under flowing conditions. *Fuel*, *69*, 794–796.

Akbarzadeh, K., & Zougari, M. (2008). Introduction to a novel approach for modeling wax deposition in fluid flows. 1. Taylor–Couette system. *Industrial and Engineering Chemistry Research*, *47*, 953–963.

Alana, J. D. (2003). *Investigation of heavy oil single-phase paraffin deposition characteristics* (M.S. thesis). University of Tulsa, Tulsa, OK.

Alcazar-Vara, L. A., & Buenrostro-Gonzalez, E. (2011). Characterization of the wax precipitation in Mexican crude oils. *Fuel Processing Technology*, *92*, 2366–2374.

Alghanduri, L. M., Elgarni, M. M., Daridon, J.-L., & Coutinho, J. A. P. (2010). Characterization of Libyan waxy crude oils. *Energy & Fuels*, *24*, 3101–3107.

Apte, M. S., Matzain, A., Zhang, H.-Q., Volk, M., Brill, J., & Creek, J. L. (2001). Investigation of paraffin deposition during multiphase flow in pipelines and wellbores—Part 2—Modeling. *Journal of Energy Resources Technology*, *123*, 150–157.

Ashford, J. D., Blount, C. G., Marcou, J. A., & Ralph, J. M. (1990). Annular packer fluids for paraffin control: Model study and successful field application. *SPE Production Engineering*, *5*, 351–355.

ASTM International. (2008). *Standard test method for temperature calibration of differential scanning calorimeters and differential thermal analyzers* (ASTM Standard E967-08). doi:10.1520/E0967-08.

ASTM International. (2010). *Standard test method for measurement of transition temperatures of petroleum waxes by differential scanning calorimetry (DSC)* (ASTM Standard D4419-90). doi:10.1520/D4419-90R10.

ASTM International. (2011). *Standard test method for cloud point of petroleum products* (ASTM Standard D2500-11). doi:10.1520/D2500-11.

Bai, C., & Zhang, J. (2013a). Effect of carbon number distribution of wax on the yield stress of waxy oil gels. *Industrial and Engineering Chemistry Research*, *52*, 2732–2739.

Bai, C., & Zhang, J. (2013b). Thermal, macroscopic, and microscopic characteristics of wax deposits in field pipelines. *Energy & Fuels*, *27*, 752–759.

Bai, Y., & Bai, Q. (2012). *Subsea engineering handbook* (1st ed.). Houston, TX: Gulf Professional Publishing.

Barry, E. G. (1971). Pumping non-Newtonian waxy crude oils. *Journal of the Institute of Petroleum*, *57*, 74–85.

Bendiksen, K. H., Maines, D., Moe, R., & Nuland, S. (1991). The dynamic two-fluid model OLGA: Theory and application. *SPE Production Engineering*, *6*, 171–180.

Bern, P. A., Withers, V. R., & Cairns, R. J. R. (1980). Wax deposition in crude oil pipelines. In *European Offshore Petroleum Conference & Exhibition* (p. 571). London: Earls Court.

Berne-Allen, A., Jr., & Work, L. T. (1938). Solubility of refined paraffin waxes in petroleum fractions. *Industrial and Engineering Chemistry*, *30*, 806–812.

Bhat, N. V., & Mehrotra, A. K. (2004). Measurement and prediction of the phase behavior of wax–solvent mixtures: Significance of the wax disappearance temperature. *Industrial and Engineering Chemistry Research, 43*, 3451–3461.

Bidmus, H. O., & Mehrotra, A. K. (2009). Solids deposition during "cold flow" of wax–solvent mixtures in a flow-loop apparatus with heat transfer. *Energy & Fuels, 23*, 3184–3194.

Bilderback, C. A., & McDougall, L. A. (1969). Complete paraffin control in petroleum production. *Journal of Petroleum Technology, 21*, 1151–1156.

Bokin, E., Febrianti, F., Khabibullin, E., & Perez, C. E. S. (2010). *Flow assurance and sour gas in natural gas production.*

Brill, J. P. (1987). Multiphase flow in wells. *Journal of Petroleum Technology, 39*, 15–21.

Brown, T. S., Niesen, V. G., & Erickson, D. D. (1993). Measurement and prediction of the kinetics of paraffin deposition. *Proceedings—SPE Annual Technical Conference and Exhibition*, 353–368.

Bruno, A., Sarica, C., Chen, H., & Volk, M. (2008). Paraffin deposition during the flow of water-in-oil and oil-in-water dispersions in pipes. In *ACTE 2008 Proceedings: Proceedings of SPE Annual Technical Conference and Exhibition*. Denver, CO: Society of Petroleum Engineers, SPE 114747, September 21–24, 2008.

Burger, E. D., Perkins, T. K., & Striegler, J. H. (1981). Studies of wax deposition in the Trans Alaska Pipeline. *Journal of Petroleum Technology, 33*, 1075–1086.

Cazaux, G., Barre, L., & Brucy, F. (1998). Waxy crude cold start: Assessment through gel structural properties. In *SPE Annual Technical Conference and Exhibition*. New Orleans, LA: Society of Petroleum Engineers, SPE 49213.

Chilton, T. H., & Colburn, A. P. (1934). Mass transfer (absorption) coefficients prediction from data on heat transfer and fluid friction. *Industrial and Engineering Chemistry, 26*, 1183–1187.

Chueh, P. L., & Prausnitz, J. M. (1967). Vapor–liquid equilibria at high pressures: Calculation of partial molar volumes in nonpolar liquid mixtures. *AIChE Journal, 13*, 1099–1107.

Claudy, P., Letoffe, J. M., Neff, B., & Damin, B. (1986). Diesel fuels: Determination of onset crystallization temperature, pour point and filter plugging point by differential scanning calorimetry. Correlation with standard test methods. *Fuel, 65*, 861–864.

Cleaver, J., & Yates, B. (1973). Mechanism of detachment of colloidal particles from a flat substrate in a turbulent flow. *Journal of Colloid and Interface Science, 44*, 464–474.

Coto, B., Coutinho, J. A. P., Martos, C., Robustillo, M. D., Espada, J. J., & Peña, J. L. (2011). Assessment and improvement of n-paraffin distribution obtained by HTGC to predict accurately crude oil cold properties. *Energy & Fuels, 25*, 1153–1160.

Coto, B., Martos, C., Espada, J. J., Robustillo, M. D., Merino-García, D., & Pe, L. (2011). A new DSC-based method to determine the wax porosity of mixtures precipitated from crude oils. *Energy & Fuels, 25*, 1707–1713.

Coto, B., Martos, C., Espada, J. J., Robustillo, M. D., & Peña, J. L. (2010). Analysis of paraffin precipitation from petroleum mixtures by means of DSC: Iterative procedure considering solid–liquid equilibrium equations. *Fuel, 89*, 1087–1094.

Coto, B., Martos, C., Espada, J. J., Robustillo, M. D., Peña, J. L., Coutinho, J. A. P., & Pe, L. (2011). Study of new methods to obtain the n-paraffin distribution of crude oils and its application to flow assurance. *Energy & Fuels, 25*, 487–492.

Coto, B., Martos, C., Peña, J. L., Espada, J. J., & Robustillo, M. D. (2008). A new method for the determination of wax precipitation from non-diluted crude oils by fractional precipitation. *Fuel*, *87*, 2090–2094.

Coutinho, J. A. P. (1998). Predictive UNIQUAC: A new model for the description of multiphase solid–liquid equilibria in complex hydrocarbon. *Industrial and Engineering Chemistry Research*, *37*, 4870–4875.

Coutinho, J. A. P., & Daridon, J. L. (2001). Low-pressure modeling of wax formation in crude oils. *Energy & Fuels*, *15*, 1454–1460.

Coutinho, J. A. P., & Daridon, J.-L. (2005). The limitations of the cloud point measurement techniques and the influence of the oil composition on its detection. *Petroleum Science and Technology*, *23*, 1113–1128.

Coutinho, J. A. P., Edmonds, B., Moorwood, T., Szczepanski, R., & Zhang, X. (2006). Reliable wax predictions for flow assurance. *Energy & Fuels*, *20*, 1081–1088.

Coutinho, J. A. P., & Ruffier-Méray, V. (1997). Experimental measurements and thermodynamic modeling of paraffinic wax formation in undercooled solutions. *Industrial and Engineering Chemistry Research*, *36*, 4977–4983.

Coutinho, J. A. P., & Stenby, E. H. (1996). Predictive local composition models for solid/liquid equilibrium in n-alkane systems: Wilson equation for multicomponent systems *Industrial and Engineering Chemistry Research*, *35*, 918–925.

Coutinho, P. (1999). Predictive local composition models: NRTL and UNIQUAC and their application to model solid–liquid equilibrium of n-alkanes. *Fluid Phase Equilibria*, *158*, 447–457.

Couto, G. (2004). *Investigation of two-phase oil-water paraffin deposition* (M.S. thesis). University of Tulsa, Tulsa, OK.

Creek, J., Lund, H. J., Brill, J. P., & Volk, M. (1999). Wax deposition in single phase flow. *Fluid Phase Equilibria*, *158–160*, 801–811.

Crochet, M. J., Davies, A. R., & Walters, K. (1991). *Numerical simulation of non-Newtonian flow* (3rd ed.). New York: Elsevier B.V.

Daridon, J., Coutinho, J. A. P., & Montel, F. (2001). Solid-liquid-vapor phase boundary of a North Sea waxy crude: Measurement and modeling. *Energy & Fuels*, *15*, 730–735.

Dauphin, C., Daridon, J., Coutinho, J., Baylère, P., & Potin-Gautier, M. (1999). Wax content measurements in partially frozen paraffinic systems. *Fluid Phase Equilibria*, *161*, 135–151.

Deen, W. M. (1998). *Analysis of transport phenomena* (2nd ed.). Oxford University Press.

DeepStar. (2011). DeepStar 10203 wax prediction confidence and pigging risk. Retrieved from http://www.deepstar.org/PhaseXProjects/.

de Oliveira, M. C. K., Teixeira, A., Vieira, L. C., de Carvalho, R. M., de Carvalho, A. B. M., & do Couto, B. C. (2012). Flow assurance study for waxy crude oils. *Energy & Fuels*, *26*, 2688–2695.

Ding, J., Zhang, J., Li, H., Zhang, F., & Yang, X. (2006). Flow behavior of Daqing waxy crude oil under simulated pipelining conditions. *Energy & Fuels*, *20*, 2531–2536.

Dirand, M., Bouroukba, M., Briard, A.-J., Chevallier, V., Petitjean, D., & Corriou, J.-P. (2002). Temperatures and enthalpies of (solid + solid) and (solid + liquid) transitions of n-alkanes. *The Journal of Chemical Thermodynamics*, *34*, 1255–1277.

Dirand, M., Chevallier, V., & Provost, E. (1998). Multicomponent paraffin waxes and petroleum solid deposits: Structural and thermodynamic state. *Fuel*, *77*, 1253–1260.

Dittus, F., & Boelter, L. (1985). Heat transfer in automobile radiators of the tubular type. *International Communications in Heat and Mass Transfer, 12*, 3–22.

Eaton, P. E., & Weeter, G. Y. (1976). Paraffin deposition in flow lines. In *16th National Heat Transfer Conference*. St. Louis, MO.

Economic analysis methodology for the 5-year OCS Oil and Gas Leasing Program for 2012–2017. (2011). Washington, DC.

Edmonds, B., Moorwood, T., Szczepanski, R., & Zhang, X. (2007). Simulating wax deposition in pipelines for flow assurance. *Energy & Fuels, 22*, 729–741.

Elliott, J. R., & Lira, C. T. (2012). *Introductory chemical engineering thermodynamics*. Upper Saddle River, NJ: Prentice Hall.

Elsharkawy, A. M., Al-Sahhaf, T. A., & Fahim, M. A. (2000). Wax deposition from Middle East crudes. *Fuel, 79*, 1047–1055.

Erickson, D. D., Niesen, V. G., & Brown, T. S. (1993). Thermodynamic measurement and prediction of paraffin precipitation in crude oil. In *SPE Annual Technical Conference and Exhibition* (pp. 353–368). Houston, TX: Society of Petroleum Engineers.

Esbensen, K. H., Halstensen, M., Tønnesen Lied, T., Svalestuen, J., de Silva, S., & Hope, B. (1998). Acoustic chemometrics—From noise to information. *Chemometrics and Intelligent Laboratory Systems, 44*, 61–76.

Garcia, M. C. (2001). Paraffin deposition in oil production. In *SPE International Symposium on Oilfield Chemistry* (pp. 1–7). Houston, TX: Society of Petroleum Engineers.

Garcia, M., Lopez, F., & Nino, Y. (1995). Characterization of near-bed coherent structures in turbulent open channel flow using synchronized high-speed video and hot-film measurements. *Experiments in Fluids, 19*, 16–28.

Gluyas, J. G., & Underhill, J. R. (2003). The Staffa Field, Block 3/8b, UK North Sea. *Geological Society, London, Memoirs, 20*, 327–333.

Gnielinski, V. (1976). New equations for heat and mass transfer in turbulent pipe and channel flow. *International Journal of Chemical Engineering, 16*, 359–368.

Golczynski, T. S., & Kempton, E. (2006). Understanding wax problems leads to deep-water flow assurance solutions. *World Oil*, D–7–D–10.

Green, D. W. (2008). *Perry's chemical engineers' handbook* (8th ed.). New York: McGraw-Hill.

Haaland, S. E. (1983). Simple and explicit formulas for the friction factor in turbulent pipe flow. *Journal of Fluids Engineering, 105*, 89–90.

Halstensen, M., Arvoh, B. K., Amundsen, L., & Hoffmann, R. (2013). Online estimation of wax deposition thickness in single-phase sub-sea pipelines based on acoustic chemometrics: A feasibility study. *Fuel, 105*, 718–727.

Hammami, A., & Mehrotra, A. K. (1995). Liquid-solid-solid thermal behaviour of n-C44+n-C50 and n-C25+n-C28 paraffinic binary mixtures. *Fluid Phase Equilibria, 111*, 253–272.

Han, S., Huang, Z., Senra, M., Hoffmann, R., & Fogler, H. S. (2010). Method to determine the wax solubility curve in crude oil from centrifugation and high temperature gas chromatography measurements. *Energy & Fuels, 24*, 1753–1761.

Hansen, A. B., Larsen, E., Pedersen, W. B., Nielsen, A. B., & Rønningsen, H. P. (1991). Wax precipitation from North Sea crude oils. 3. Precipitation and dissolution of wax studied by differential scanning calorimetry. *Energy & Fuels, 5*, 914–923.

Hansen, J. H., Fredenslund, A., Pedersen, K. S., & Røningsen, H. P. (1988). A thermodynamic model for predicting wax formation in crude oils. *AIChE Journal, 34*, 1937–1942.

Hayduk, W., & Minhas, B. (1982). Correlations for prediction of molecular diffusivities in liquids. *Canadian Journal of Chemical Engineering, 60*, 295–299.

Hernandez, O. C. (2002). *Investigation of single-phase paraffin deposition characteristics* (M.S. thesis). University of Tulsa, Tulsa, OK.

Hernandez, O. C., Hensley, H., Sarica, C., Brill, J. P., Volk, M., & Delle-Case, E. (2003). Improvements in single-phase paraffin deposition modeling. In *SPE Annual Technical Conference and Exhibition* (pp. 1–9). Denver, CO: Society of Petroleum Engineers.

Hoffmann, R., & Amundsen, L. (2010). Single-phase wax deposition experiments. *Energy & Fuels, 24*, 1069–1080.

Hsu, J. J. C., & Brubaker, J. P. (1995). Wax deposition measurement and scale-up modeling for waxy live crudes under turbulent flow conditions. In *International Meeting on Petroleum Engineering* (pp. 1–10). Beijing, China: Society of Petroleum Engineers.

Huang, Z., Lee, H. S., Senra, M., & Fogler, H. S. (2011). A fundamental model of wax deposition in subsea oil pipelines. *AIChE Journal, 57*, 2955–2964.

Huang, Z., Lu, Y., Hoffmann, R., Amundsen, L., & Fogler, H. S. (2011). The effect of operating temperatures on wax deposition. *Energy & Fuels, 25*, 5180–5188.

Huang, Z., Senra, M., Kapoor, R., & Fogler, H. S. (2011). Wax deposition modeling of oil/water stratified channel flow. *AIChE Journal, 57*, 841–851.

Hunt, E. B., Jr. (1962). Laboratory study of paraffin deposition. *Journal of Petroleum Technology, 14*, 1259–1269.

Ijeomah, C. E., Dandekar, A. Y., Chukwu, G. A., Khataniar, S., Patil, S. L., & Baldwin, A. L. (2008). Measurement of wax appearance temperature under simulated pipeline (dynamic) conditions. *Energy & Fuels, 22*, 2437–2442.

Ismail, L., Westacott, R. E., & Ni, X. (2008). On the effect of wax content on paraffin wax deposition in a batch oscillatory baffled tube apparatus. *Chemical Engineering Journal, 137*, 205–213.

Jemmett, M. R., Deo, M., Earl, J., & Mogenhan, P. (2012). Applicability of cloud point depression to "cold flow." *Energy & Fuels, 26*, 2641–2647.

Jennings, D. W., & Weispfennig, K. (2005). Effects of shear and temperature on wax deposition: Coldfinger investigation with a Gulf of Mexico crude oil. *Energy & Fuels, 19*, 1376–1386.

Jiang, Z., Hutchinson, J., & Imrie, C. (2001). Measurement of the wax appearance temperatures of crude oils by temperature modulated differential scanning calorimetry. *Fuel, 80*, 367–371.

Jimenez, J., Moin, P., Moser, R., & Keefe, L. (1988). Ejection mechanisms in the sublayer of a turbulent channel. *Physics of Fluids, 31*, 1311–1313.

Juyal, P., Cao, T., Yen, A., & Venkatesan, R. (2011). Study of live oil wax precipitation with high-pressure micro-differential scanning calorimetry. *Energy & Fuels, 25*, 568–572.

Karacan, C. O., Demiral, M. R. B., & Kok, M. V. (2000). Application of x-ray CT imaging as an alternative tool for cloud point determination. *Petroleum Science and Technology, 18*, 835–849.

Kaya, A. S., Sarica, C., & Brill, J. P. (1999). Comprehensive mechanistic modeling of two-phase flow in deviated wells. In *SPE Annual Technical Conference and Exhibition*. Houston, TX: Society of Petroleum Engineers.

Kim, D., Ghajar, A. J., Dougherty, R. L., & Ryali, V. K. (1999). Comparison of twenty two-phase heat transfer correlations with seven sets of experimental data, including flow pattern and tube inclination effects. *Heat Transfer Engineering, 20*(1), 15–40.

Kleinhans, J., Niesen, V., & Brown, T. (2000). Pompano paraffin calibration field trials. In *SPE Annual Technical Conference and Exhibition* (pp. 1–15). Dallas, TX: Society of Petroleum Engineers.

Kok, M. V., Jean-Marie, L., Claudy, P., Martin, D., Garcin, M., Vollet, J., … Ura, C. (1996). Comparison of wax appearance temperatures of crude oils by DSC, thermomicroscopy and viscometry. *Fuel, 75,* 787–790.

Kok, M. V., Letoffe, J. M., & Claudy, P. (1999). DSC and rheometry investigations of crude oils. *Journal of Thermal Analysis and Calorimetry, 56,* 959–965.

Kravchenko, V. (1946). No title. *Acta Physicochimica, 21,* 335.

Kruka, V. R., Cadena, E. R., & Long, T. E. (1995). Cloud-point determination for crude oils. *Journal of Petroleum Technology, 47,* 681–687.

Labes-Carrier, C., Rønningsen, H. P., Kolnes, J., & Leporcher, E. (2002). Wax deposition in North Sea gas condensate and oil systems: Comparison between operational experience and model prediction. In *SPE Annual Technical Conference and Exhibition.* San Antonio, TX.

Lee, H. S. (2008). *Computational and rheological study of wax deposition and gelation in subsea pipelines* (Ph.D. thesis). University of Michigan.

Leontaritis, K. J. (1996). The asphaltene and wax deposition envelopes. *Fuel Science and Technology International, 14,* 13–39.

Li, H., & Zhang, J. (2003). A generalized model for predicting non-Newtonian viscosity of waxy crudes as a function of temperature and precipitated wax. *Fuel, 82,* 1387–1397.

Lindeloff, N., & Krejbjerg, K. (2002). A compositional model simulating wax deposition in pipeline systems. *Energy & Fuels, 16,* 887–891.

Lira-Galeana, C., Firoozabadi, A., & Prausnitz, J. M. (1996). Thermodynamics of wax precipitation in petroleum mixtures. *AIChE Journal, 42,* 239–248.

Lu, Y., Huang, Z., Hoffmann, R., Amundsen, L., Fogler, H. S., & Sheng, Z. (2012). Counterintuitive effects of the oil flow rate on wax deposition. *Energy & Fuels, 26,* 4091–4097.

Lund, H.-J. (1998). *Investigation of paraffin deposition during single-phase liquid flow in pipelines* (M.S. thesis). University of Tulsa.

Martos, C., Coto, B., Espada, J. J., Robustillo, M. D., Gómez, S., & Peña, J. L. (2008). Experimental determination and characterization of wax fractions precipitated as a function of temperature. *Energy & Fuels, 22,* 708–714.

Martos, C., Coto, B., Espada, J. J., Robustillo, M. D., Peña, J. L., & Merino-Garcia, D. (2010). Characterization of Brazilian crude oil samples to improve the prediction of wax precipitation in flow assurance problems. *Energy & Fuels, 24,* 2221–2226.

Matzain, A. (1997). *Single phase liquid paraffin deposition modeling* (M.S. thesis). University of Tulsa.

Matzain, A., Apte, M. S., Zhang, H.-Q., Volk, M., Redus, C. L., Brill, J. P., & Creek, J. L. (2001). Multiphase flow wax deposition modeling. In *ETCE 2001: Petroleum Production Technology Symposium.* Houston, TX.

Merino-Garcia, D., & Correra, S. (2008). Cold flow: A review of a technology to avoid wax deposition. *Petroleum Science and Technology, 26,* 446–459.

Monger-McClure, T. G., Tackett, J. E., & Merrill, L. S. (1999). Comparisons of cloud point measurement and paraffin prediction methods. *SPE Production & Facilities, 14,* 4–16.

Monrad, C. C., & Pelton, J. G. (1942). Heat transfer by convection in annular spaces. *Transations of AIChE, 38,* 593–611.

Mukherjee, H., & Brill, J. P. (1985). Pressure drop correlations for inclined two-phase flow. *Journal of Energy Resources Technology, 107*, 549–554.

Musser, B. J., Kilpatrick, P. K., & Carolina, N. (1998). Molecular characterization of wax isolated from a variety of crude oils. *Energy & Fuels, 59*, 715–725.

Nazar, A. R. S., Dabir, B., Vaziri, H., & Islam, M. R. (2001). Experimental and mathematical modeling of wax deposition and propagation in pipes transporting crude oil. In *SPE Production and Operations Symposium* (pp. 1–11). Oklahoma City, OK: Society of Petroleum Engineers.

Niesen, V. (2002). The real cost of subsea pigging. *E&P Magazine*, 97–98.

Noville, I., & Naveira, L. (2012). Comparison between real field data and the results of wax deposition simulation. In *SPE Latin America and Caribbean Petroleum Engineering Conference* (pp. 1–12). Mexico City, Mexico: Society of Petroleum Engineers.

Paso, K. G., & Fogler, H. S. (2004). Bulk stabilization in wax deposition systems. *Energy & Fuels, 18*, 1005–1013.

Paso, K., Kallevik, H., & Sjöblom, J. (2009). Measurement of wax appearance temperature using near-infrared (NIR) scattering. *Energy & Fuels, 23*, 4988–4994.

Pauly, J., Daridon, J.-L., & Coutinho, J. A. P. (2004). Solid deposition as a function of temperature in the nC10 + (nC24–nC25–nC26) system. *Fluid Phase Equilibria, 224*, 237–244.

Pauly, J., Dauphin, C., & Daridon, J. L. (1998). Liquid–solid equilibria in a decane+multi-paraffins system. *Fluid Phase Equilibria, 149*, 191–207.

Pedersen, K. S., Skovborg, P., & Rønningsen, H. P. (1991). Wax precipitation from North Sea crude oils. 4. Thermodynamic modeling. *Energy & Fuels, 5*, 924–932.

Pedersen, K. S., Thomassen, P., & Fredenslund, A. (1985). Thermodynamics of petroleum mixtures containing heavy hydrocarbons. 3. Efficient flash calculation procedures using the SRK equation of state. *Industrial & Engineering Chemistry Process Design and Development, 24*, 948–954.

Pedersen, W. B., Hansen, A. B., Larsen, E., Nielsen, A. B., & Rønningsen, H. P. (1991). Wax precipitation from North Sea crude oils. 2. Solid-phase content as function of temperature determined by pulsed NMR. *Energy & Fuels, 5*, 908–913.

Petitjean, D., Schmitt, J. F., Laine, V., Bouroukba, M., Cunat, C., & Dirand, M. (2008). Presence of isoalkanes in waxes and their influence on their physical properties. *Energy & Fuels, 22*, 697–701.

Phillips, D. A., Forsdyke, I. N., McCracken, I. R., & Ravenscroft, P. D. (2011). Novel approaches to waxy crude restart: Part 2: An investigation of flow events following shut down. *J. Pet. Sci. Eng., 77*, 286–304.

PVTsim 19 method documentation. (2009). Calsep.

Reistle, C. E. (1928). *Methods of dealing with paraffin troubles encountered in producing crude oil*.

Reistle, C. E. (1932). *Paraffin and gongealing-oil problems*.

Roehner, R., & Fletcher, J. (2002). Comparative compositional study of crude oil solids from the Trans Alaska Pipeline System using high-temperature gas chromatography. *Energy & Fuels, 16*, 211–217.

Roehner, R. M., & Hanson, F. V. (2001). Determination of wax precipitation temperature and amount of precipitated solid wax versus temperature for crude oils using FT-IR spectroscopy. *Energy & Fuels, 15*, 756–763.

Rønningsen, H. P. (2012). Production of waxy oils on the Norwegian continental shelf: Experiences, challenges, and practices. *Energy & Fuels, 26*, 4124–4136.

Rønningsen, H. P., Bjamdal, B., Hansen, A. B., & Pedersen, W. B. (1991). Wax precipitation from North Sea crude oils. 1. Crystallization and dissolution temperatures, and Newtonian and non-Newtonian flow properties. *Energy & Fuels*, 5, 895–908.

RPSEA. (2007). RPSEA 1201 controlling wax deposition in the presence of hydrates technology evaluation. Retrieved from http://www.netl.doe.gov/kmd/RPSEA _Project_Outreach/07121-1201-University_of_Utah_Phase1.pdf.

Rygg, O. B., Rydahl, A. K., & Rønningsen, H. P. (1998). Wax deposition in offshore pipeline systems. In *BHRGroup 1st North American Conference Multiphase Technology*. Baniff, Canada.

Saffman, P. G. (1965). The lift on a small sphere in a slow shear flow. *Journal of Fluid Mechanics*, 22, 385–400.

Sandler, S. I. (2006). *Chemical, biochemical, and engineering thermodynamics*. Hoboken, NJ: Wiley.

Senra, M. J. (2009). *Assessing the role of polydispersity and cocrystallization on crystallizing n-alkanes in n-alkane solutions* (Ph.D. thesis). University of Michigan, Hoboken, NJ.

Sieder, E. N., & Tate, G. E. (1936). Heat transfer and pressure drop of liquids in tubes. *Industrial and Engineering Chemistry*, 28, 1429–1435.

Singh, A., Lee, H., Singh, P., & Sarica, C. (2011). Flow assurance: Validation of wax deposition models using field data from a subsea pipeline. In *Offshore Technology Conference* (pp. 1–19). Houston, TX: Offshore Technology Conference.

Singh, P. (2000). *Gel deposition on cold surfaces* (Ph.D. thesis). University of Michigan.

Singh, P., & Venkatesan, R. (2001). Morphological evolution of thick wax deposits during aging. *AIChE Journal*, 47, 6–18.

Singh, P., Venkatesan, R., Fogler, H. S., & Nagarajan, N. (2000). Formation and aging of incipient thin film wax-oil gels. *AIChE Journal*, 46, 1059–1074.

Smith, B. (1999). *Infrared spectral interpretation, a systematic approach*. New York: CRC Press.

Snyder, R. G., Conti, G., Strauss, H. L., & Dorset, D. L. (1993). Thermally induced mixing in partially microphase segregated binary n-alkane crystals. *The Journal of Physical Chemistry*, 97, 7342–7350.

Snyder, R. G., Goh, M. C., Srivatsavoy, V. J. P., Strauss, H. L., & Dorset, D. L. (1992). Measurement of the growth kinetics of microdomains in binary n-alkane solid solutions by infrared spectroscopy. *The Journal of Physical Chemistry*, 96, 10008–10019.

Snyder, R. G., Hallmark, V. M., Strauss, H. L., & Maroncelli, M. (1986). Temperature and phase behavior of infrared intensities: The poly(methylene) chain. *The Journal of Physical Chemistry*, 94720, 5623–5630.

Snyder, R. G., Srivatsavoy, V. J. P., Cates, D. A., Strauss, H. L., White, J. W., & Dorset, D. L. (1994). Hydrogen/deuterium isotope effects on microphase separation in unstable crystalline mixtures of binary n-alkanes. *The Journal of Physical Chemistry*, 98, 674–684.

Streitwieser, A., & Heathcock, C. (1976). *Introduction to organic chemistry*. New York: Macmillan Publishing.

Suppiah, S., Ahmad, A., Alderson, C., Akbarzadeh, K., Gao, J., Shorthouse, J., ... Jamaluddin, A. (2010). Waxy crude production management in a deepwater subsea environment. In *SPE Annual Technical Conference and Exhibition* (pp. 1–18). Florence, Italy: Society of Petroleum Engineers.

Urushihara, T., Meinhart, C. D., & Adrain, R. J. (1993). Investigation of the logarithmic layer in pipe flow using particle image velocimetry. In *Near Wall Turbulent Flows*, (pp. 433–446). Amsterdam, Netherlands: Elsevier.

Van Driest, E. R. (1956). On turbulent flow near a wall. *J. Aeronaut. Sci.*, *23*, 1007–1011.

Venkatesan, R. (2004). *The deposition and rheology of organic gels* (Ph.D. thesis). University of Michigan.

Venkatesan, R., & Fogler, H. S. (2004). Comments on analogies for correlated heat and mass transfer in turbulent flow. *AIChE Journal, 50*(7), 1623–1626.

Venkatesan, R., Nagarajan, N. R., Paso, K., Yi, Y.-B., Sastry, A. M., & Fogler, H. S. (2005). The strength of paraffin gels formed under static and flow conditions. *Chemical Engineering Science, 60*, 3587–3598.

Vieira, L. C., Buchuid, M. B., & Lucas, E. F. (2010). Effect of pressure on the crystallization of crude oil waxes. I. Selection of test conditions by microcalorimetry. *Energy & Fuels, 24*, 2208–2212.

Wilke, C. R., & Chang, P. (1955). Correlation of diffusion coefficients in dilute solutions. *AIChE Journal, 1*, 264–270.

Wilkes, J. O. (2005). *Fluid mechanics for chemical engineers* (2nd ed.). Upper Saddle River, NJ: Prentice Hall.

Won, K. W. (1986). Thermodynamics for solid solution-liquid-vapor equilibria: Wax phase formation from heavy hydrocarbon mixtures. *Fluid Phase Equilibria, 30*, 265–279.

Xiao, J. J., Shonham, O., & Brill, J. P. (1990). Comprehensive mechanistic model for two-phase flow in pipelines. In *SPE Annual Technical Conference and Exhibition*, New Orleans, LA: Society of Petroleum Engineers.

Yan, D., & Luo, Z. (1987). Rheological properties of Daqing crude oil and their application in pipeline transportation. *SPE Production Engineering, 2*, 267–276.

Zheng, S., Zhang, F., Huang, Z., & Fogler, H. S. (2013). Effects of operating conditions on wax deposit carbon number distribution: Theory and experiment. *Energy & Fuels, 27*, 7379–7388.

Zhu, T., Walker, J. A., & Liang, J. (2008). *Evaluation of wax deposition and its control during production of Alaska North Slope Oils—Final Report*.

Appendix A: Nomenclature

A_{HM}: coefficient in the Hayduk–Minhas correlation for wax mass diffusivity

B_{WC}: coefficient in the Wilke–Chang correlation for wax mass diffusivity

C: concentration of dissolved waxy components (kg/m^3)

$C(eq)$: concentration of the dissolved waxy components based on thermo-dynamic equilibrium (kg/m^3)

C_{inlet}: concentration of dissolved waxy components at the inlet of the pipe (kg/m^3)

C_{oil}: concentration of dissolved waxy components at the bulk oil (kg/m^3)

C_{wall}: concentration of dissolved waxy components at the pipe wall (kg/m^3)

C_p: heat capacity of the oil (J/mol/K)

D_{wax}: mass diffusivity (diffusion coefficient) of wax in oil (m^2/s)

$D_{wax,interface}$: mass diffusivity (diffusion coefficient) of wax in oil based on the temperature at the oil–deposit interface (m^2/s)

$D_{wax,wall}$: mass diffusivity (diffusion coefficient) of wax in oil based on the wall temperature (m^2/s)

ΔCp_i: change of heat capacity of component i due to phase transition (J/mol/K)

F_{wax}: mass fraction of wax in the deposit

F_i: mass fraction of a certain waxy component in the deposit

G^E: excess Gibbs free energy due to molecular interactions (J)

ΔG_{mix}: change of Gibbs free energy due to mixing (J)

ΔH_i^f: heat of fusion of component i (J/mol)

ΔH_i^{Tr}: heat of solid–solid phase transition of component i (J/mol)

ΔH_i^{Sub}: heat of sublimation of component i (J/mol)

J_A: mass flux of waxy components from the bulk oil toward the deposit surface (kg/m^2/s)

J_B: mass flux of waxy components into the deposit (kg/m^2/s)

J_{wax}: characteristic mass flux of wax deposition (kg/m^2/s)

L_{pipe}: length of the pipe (m)

$L_{pipe,i}$: length of the ith section of the pipe (m)

M_B: solvent molecular weight in the Hayduk–Minhas correlation for the wax diffusivity (g/mol)

ΔP_{pipe}: pressure drop across the pipe (Pa)

$\Delta P_{pipe,i}$: pressure drop across the ith section of the pipe (Pa)

Q_{oil}: volumetric flow rate of oil (m^3/s)

$\Delta Q_{thermal}$: difference in the thermal energy flow rate between the inlet and the outlet (J/s)

R: gas constant (J/mol/K)

$R_{interface}$: effective radius of the oil flow during wax deposition, i.e., the distance from the pipe centerline to the oil–deposit interface (m)

R_{pipe}: radius of the pipe (m)

T: temperature (K)

$T_{ambient}$: ambient temperature (K)

$T_{coolant}$: bulk temperature of the coolant (K)

$T_{coolant,inlet}$: temperature of the coolant at the inlet of the coolant pipe (K)

$T_{coolant,outlet}$: temperature of the coolant at the outlet of the coolant pipe (K)

$T_{interface}$: temperature at the oil–deposit interface (K)

T_i^f: melting point of component i (K)

T_i^p: precipitation temperature of component i (K)

T_i^{Tr}: temperature of solid–solid phase transition of component i (K)

T_{oil}: temperature of the bulk oil (K)

$T_{oil,inlet}$: temperature of the oil at the inlet of the pipe (K)

$T_{oil,outlet}$: temperature of the oil at the outlet of the pipe (K)

T_{wall}: temperature at the pipe wall (K)

ΔT_{lm}: logarithmic mean temperature difference for heat exchangers (K)

U: average velocity of the oil in the axial direction (m/s)

$U_{overall}$: overall heat transfer coefficient of the pipe in the radial direction based on the outer diameter of the pipe (W/m²/K)

V_A: averaged molar volume of n-paraffins (cm³/mol)

V_i: molar volume of component i (L/mol)

V_{wi}: van der Waals volume of component i (L/mol)

ΔV_i: change of molar volume of component i at phase transition (L/mol)

V_z: velocity of oil in the axial direction (m)

Z_L: equation of state compressibility factor of the liquid phase

Z_V: equation of state compressibility factor of the vapor phase

$a_{mixture}$: equation of state parameter (N × m⁴)

$b_{mixture}$: equation of state parameter (m³)

$d_{effective\ inner}$: effective flowing inner diameter of the pipe accounting for the thickness of the wax deposit (m)

d_{inner}: inner diameter of the pipe (m)

d_{outer}: outer diameter of the pipe (m)

d_{pipe}: diameter of the pipe (m)

f_{Darcy}: Darcy friction factor

f_i^L: fugacity of component i in the liquid phase (Pa)

f_i^S: fugacity of component i in the solid phase (Pa)

f_i^V: fugacity of component i in the vapor phase (Pa)

$h_{coolant}$: heat transfer coefficient of the coolant in a flow-loop wax deposition experiment (W/m²/K)

$h_{internal}$: internal heat transfer coefficient based on the pipe inner diameter (W/m²/K)

$h_{external}$: external heat transfer coefficient (including the pipe wall, pipe insulation, and the ambient environment) based on the pipe inner diameter (W/m²/K)

$k_{mass\ transfer}$: mass transfer coefficient of wax in the bulk oil (m/s)

$k_{deposit}$: thermal conductivity of wax deposit (W/m/K)

k_{oil}: thermal conductivity of oil (W/m/K)

k_{pipe}: thermal conductivity of pipe (W/m/K)

$k_{precipitation}$: precipitation rate constant in the Michigan Wax Predictor wax deposition model (s^{-1})

n_i^F: number of moles of the component i in the feed (mol)

n^L: number of moles of the liquid phase (mol)

n^S: number of moles of the solid phase (mol)

n^V: number of moles of the vapor phase (mol)

p: pressure (Pa)

q: measured heat released due to crystallization (J/mol)

r: radial coordinate (m)

S_i: mole fraction of component i in the solid phase

t: time (s)

w_i: amount of component i precipitated (mol)

x_i: mole fraction of component i in the liquid phase

y_i: mole fraction of component i in the vapor phase

y^+: dimensionless distance to the wall (in turbulent fluid mechanics)

z: axial coordinate (m)

α: Wilson model binary correction factor

ε_{mass}: eddy mass diffusivity (m^2/s)

ε_{pipe}: roughness of the pipe (m)

$\varepsilon_{thermal}$: eddy thermal diffusivity (m^2/s)

η: dedimensionalized radial length of the pipe

θ: dedimensionalized concentration of waxy components

γ_i^L: activity coefficient of component i in the liquid phase

γ_i^S: activity coefficient of component i in the solid phase

δ: average solubility parameter of the phase (J$^{0.5}$/mol$^{0.5}$/L$^{0.5}$)

δ_i: solubility parameter of component i (J$^{0.5}$/mol$^{0.5}$/L$^{0.5}$)

$\delta_{deposit}$: thickness of the wax deposit (m)

$\delta_{diffusion}$: diffusion equivalent mass transfer layer (m)

$\delta_{mass\ transfer}$: thickness of the mass transfer layer (m)

$\delta_{thermal}$: conduction-equivalent boundary layer (m)

λ: dedimensionalized length of the pipe

$\lambda_{i,j}$: interaction energy between components i and j in the solid phase (J/mol)

$\mu_{centerline}$: viscosity of the pipe based on the pipe centerline temperature (Pa · s)

$\mu_{deposit\ interface}$: viscosity of the pipe based on the oil–deposit interface temperature (Pa · s)

μ_B: solvent viscosity in the Hayduk–Minhas correlation for the wax mass diffusivity (cP)

υ: dedimensionalized axial velocity of the oil

ρ_{oil}: density of the oil (kg/m^3)

$\rho_{deposit}$: density of the wax deposit (kg/m^3)

ϕ_B: association parameter for the solvent in the Wilke–Chang correlation for wax mass diffusivity

$\phi_{deposit}$: porosity (oil volume fraction) of the wax deposit

ϕ_i: composition fraction of component i defined according to the Flory free
      volume theory
$\chi_{i,j}$: Flory's binary interaction parameter between components i and j
Gz: Graetz number
Le: Lewis number
Nu: Nusselt number
Pr: Prandtl number
Re: Reynolds number
Sc: Schmidt number
Sh: Sherwood number

Index

Page numbers followed by f and t indicate figures and tables, respectively.

Printed and bound by CPI Group (UK) Ltd, Croydon, CR0 4YY

01/11/2024

01782623-0018